INTERNET THEORY, TECHNOLOGY AND APPLICATIONS

EXPLORATIONS OF INTERNET USAGE AND ACCESS IN U.S. HOMES

INTERNET THEORY, TECHNOLOGY AND APPLICATIONS

Additional books in this series can be found on Nova's website under the Series tab.

Additional E-books in this series can be found on Nova's website under the e-books tab.

COMPUTER SCIENCE, TECHNOLOGY AND APPLICATIONS

Additional books in this series can be found on Nova's website under the Series tab.

Additional E-books in this series can be found on Nova's website under the e-books tab.

INTERNET THEORY, TECHNOLOGY AND APPLICATIONS

EXPLORATIONS OF INTERNET USAGE AND ACCESS IN U.S. HOMES

BERNARD L. FOSTER
AND
MARCELINO E. GONZALES
EDITORS

Nova Science Publishers, Inc.
New York

For permission to use material from this book please contact us:
Telephone 631-231-7269; Fax 631-231-8175
Web Site: http://www.novapublishers.com

Library of Congress Cataloging-in-Publication Data

ISBN: 978-1-61942-923-9

Published by Nova Science Publishers, Inc. † New York

CONTENTS

PREFACE

The Internet is an extraordinary platform for innovation, economic growth, and social communication. High-speed Internet services delivered over broadband networks are critical to maintaining the United States' competitiveness in a global economy. A strong correlation exists between broadband and indices of economic growth, such as GDP, employment, and property values. The Administration recognizes the importance of broadband to improve health care, enhance education, and provide essential job training and employment assistance for Americans seeking work. This book examines the digital nation of computer and internet use at home with a focus on the disparities in broadband Internet adoption that exist between demographic and geographic groups.

Chapter 1- This report updates and expands last year's report, *Exploring the Digital Nation: Home Broadband Internet Adoption in the United States*, based on data from the Census Bureau's most recent Current Population Survey (CPS) School Enrollment and Internet Use Supplement. The report also provides additional information augmenting the February 2011 research preview, *Digital Nation: Expanding Internet Usage*, and includes new findings on computer and Internet use in the United States. For example, we use regression analysis to help explain some of the disparities in broadband Internet adoption that exist between demographic and geographic groups. The analysis reveals that, by holding constant certain factors such as household income, education, or age, the adoption disparities may decrease significantly.

Below is a summary of our final review of the 2010 CPS results. These findings may assist policymakers as they consider ways to promote broadband deployment and adoption in the United States.[1]

Chapter 2- The "digital divide" is a term that has been used to characterize a gap between "information haves and have-nots," or in other words, between those Americans who use or have access to telecommunications and information technologies and those who do not. One important subset of the digital divide debate concerns high-speed Internet access and advanced telecommunications services, also known as broadband. Broadband is provided by a series of technologies (e.g., cable, telephone wire, fiber, satellite, wireless) that give users the ability to send and receive data at volumes and speeds far greater than traditional "dial-up" Internet access over telephone lines.

Broadband technologies are currently being deployed primarily by the private sector throughout the United States. While the numbers of new broadband subscribers continue to grow, studies and data suggest that the rate of broadband deployment in urban/suburban and high income areas are outpacing deployment in rural and low-income areas. Some policymakers, believing that disparities in broadband access across American society could have adverse economic and social consequences on those left behind, assert that the federal government should play a more active role to avoid a "digital divide" in broadband access.

In: Explorations of Internet Usage and Access ... ISBN: 978-1- 61942-923-9
Editors: B. Foster and M. Gonzales

Chapter 1

EXPLORING THE DIGITAL NATION: COMPUTER AND INTERNET USE AT HOME*

The United States Department of Commerce

FOREWORD

The Internet is an extraordinary platform for innovation, economic growth, and social communication. High-speed Internet services delivered over broadband networks are critical to maintaining the United States' competitiveness in a global economy. A strong correlation exists between broadband (both deployment and adoption) and indices of economic growth, such as increases in Gross Domestic Product, employment, and property values. The Administration recognizes the importance of broadband to improve health care, enhance education, and provide essential job training and employment assistance for Americans seeking work. President Obama recently reiterated his long-standing commitment to ensuring broadband's role in the nation's future, stating:

> I will not sacrifice the core investments that we need to grow and create jobs. We will invest in medical research. We will invest in clean energy technology. We will invest in new roads and airports and broadband

* This is an edited, reformatted and augmented version of The United States Department of Commerce publication, prepared by Economics and Statistics Administration and National Telecommunications and Information Administration, dated November 2011.

> access. We will invest in education. We will invest in job training. We will do what we need to do to compete, and we will win the future.
>
> —Remarks by President Obama on Fiscal Policy in Washington, DC, April 13, 2011

The American Recovery and Reinvestment Act of 2009 (Recovery Act) provides a down payment on needed investments to extend the nation's broadband infrastructure, expand public computer center capacity, and promote broadband adoption. The National Telecommunications and Information Administration (NTIA) provides approximately $4 billion to fund 229 projects under the Recovery Act's Broadband Technology Opportunities Program (BTOP). These investments, combined with the Department of Agriculture's Broadband Initiatives Program, total approximately $7 billion, and they are beginning to pay dividends. BTOP grantees have installed or upgraded over 18,000 miles of new broadband networks, added or upgraded more than 16,000 computer workstations, and reported over 110,000 new subscribers as of June 30, 2011. The projects are stimulating the deployment and adoption of broadband in communities across the United States and we expect that they will continue to do so for years to come. In addition, NTIA recently launched *DigitalLiteracy.gov* in partnership with nine Federal agencies to create an online resource for improving digital literacy.

Government efforts, private sector investment, and increased demand for Internet services and applications all play a key role in facilitating the steady growth in households' computer and Internet use. For its part, the Department of Commerce is a leader in analyzing broadband access and adoption in America. In October 2010, the U.S. Census Bureau within the Economics and Statistics Administration, in collaboration with NTIA, significantly expanded the Current Population Survey (CPS) to include new questions on computer and Internet use. The Census Bureau surveyed about 54,300 households, and through statistical methods extrapolated the survey results to represent 119.5 million American households.

The CPS data revealed that 68 percent of households used broadband Internet access service, up from 64 percent the previous year. Despite this improvement, demographic and geographic disparities demonstrate a persistent digital divide among certain groups. For example, rural low-income minorities' broadband adoption at home lagged significantly behind that of other groups. In addition, almost one-third of Americans are not accessing broadband service at home. The Administration has made it a priority to evaluate the effectiveness of existing programs, develop new programs as needed, and work collaboratively with industry to design new strategies to ameliorate the digital divide. Our ongoing analysis highlights that there is no simple "one size fits all" solution to resolve the adoption disparities among broadband users. The Administration will continue to encourage Congress, state and local officials, and the private sector to find ways to promote broadband deployment and adoption so that we continue to create jobs, prepare the workforce for the rapidly developing Internet economy, and increase the nation's competitiveness.

Rebecca M. Blank
Acting Deputy Secretary of Commerce and
Under Secretary for Economic Affairs
Economics and Statistics Administration

Lawrence E. Strickling
Assistant Secretary for Communications and
Information *National Telecommunications and Information Administration*

EXECUTIVE SUMMARY

This report updates and expands last year's report, *Exploring the Digital Nation: Home Broadband Internet Adoption in the United States*, based on data from the Census Bureau's most recent Current Population Survey (CPS) School Enrollment and Internet Use Supplement. The report also provides additional information augmenting the February 2011 research preview, *Digital Nation: Expanding Internet Usage*, and includes new findings on computer and Internet use in the United States. For example, we use regression analysis to help explain some of the disparities in broadband Internet adoption that exist between demographic and geographic groups. The analysis reveals that, by holding constant certain factors such as household income, education, or age, the adoption disparities may decrease significantly.

Below is a summary of our final review of the 2010 CPS results. These findings may assist policymakers as they consider ways to promote broadband deployment and adoption in the United States.[1]

SUMMARY OF THE 2010 CPS RESULTS

- As of October 2010, more than 68 percent of households used broadband Internet access service, up from 64 percent one year earlier (Section 1, Figure 1). Approximately 80 percent of households had at least one Internet user, either at home or elsewhere (Section 3.1, Figure 3).
- Cable modem (32 percent) and DSL (23 percent) ranked as the most commonly used broadband technologies (Section 3.1, Figure 3). Other technologies, including mobile broadband, fiber optics, and satellite services, accounted for a small, but growing, segment of households with broadband Internet access service.
- Dial-up use at home – the preferred form of residential Internet access through the mid 2000s – continued to decline from five percent in October 2009 to three percent one year later (Section 3.1, Figure 3).
- Over three-fourths (77 percent) of households had a computer – the principal means by which households access the Internet – compared with 62 percent in 2003 (Section 1, Figure 1). Low computer use correlates with low broadband adoption rates.
- Broadband Internet adoption, as well as computer use, varied across demographic and geographic groups. Lower income families, people with less education, those with disabilities, Blacks, Hispanics, and rural residents generally lagged the national average in both broadband adoption and computer use. For example, home broadband adoption and computer use stood at only 16 percent and 27 percent, respectively, among rural households headed by a Black householder without a high school diploma (Section 4.2, Table 4). Also, households with school-age children exhibited higher broadband adoption and computer use rates than other households (Section 4.1, Figure 7).
- The differences in socio-economic attributes do not entirely explain why some groups lagged in adoption. Broadband Internet adoption disparities decrease when regression analysis holds constant certain household characteristics, such as income, education, race, ethnicity, foreign-born status, household composition, disability status, or geographic location. For example, the gap with respect to broadband Internet adoption associated with disabilities decreases from 29 to six

percentage points when controlling for income, education, age, and other attributes (Section 4.3, Figure 18).

- The most important reasons households without broadband Internet or dial-up service gave for not subscribing were: (1) lack of need or interest (47 percent); (2) lack of affordability (24 percent); and (3) inadequate computer (15 percent) (Section 5, Figure 19).
- Households reporting affordability as the major barrier to subscribing to broadband service cited both the fixed cost of purchasing a computer and the recurring monthly subscription costs as important factors (Section 5, Figure 21). Our analysis of the expanded CPS data suggests that work, school, public libraries, and someone else's house were all popular alternatives for Internet access among those with no home broadband Internet access service (Section 6, Figure 23). Not surprisingly, individuals with no home broadband Internet access service relied on locations such as public libraries (20 percent) or other people's houses (12 percent) more frequently than those who used broadband Internet access service at home.

1. Introduction

The Internet has transformed our social and economic environment by providing an important platform for innovation, economic growth, and social communication. Residential use of broadband Internet access services has risen dramatically during the past decade, demonstrating that the Internet plays a key role in the everyday lives of many people.

Nonetheless, as Figure 1 illustrates, not everyone uses a computer or the Internet. Today, there exists a substantial gap between these technologies' adoption rates (77 and 71 percent, respectively) and that of telephone service (96 percent); broadband Internet access (68 percent) trails as well. However, the differentials have decreased significantly between very mature telephone subscribership and growing digital technologies. For example, in 2003 computer use lagged by approximately 33 percentage points and broadband by almost 75 percentage points. While telephone diffusion has plateaued for more than two decades, nascent computer and Internet adoption rates have continued to climb, albeit at a slower pace than earlier in the decade. In contrast, broadband Internet's rise has remained sharp and robust, as dial-up adoption plunged from 37 percent in 2000 to three percent in 2010.

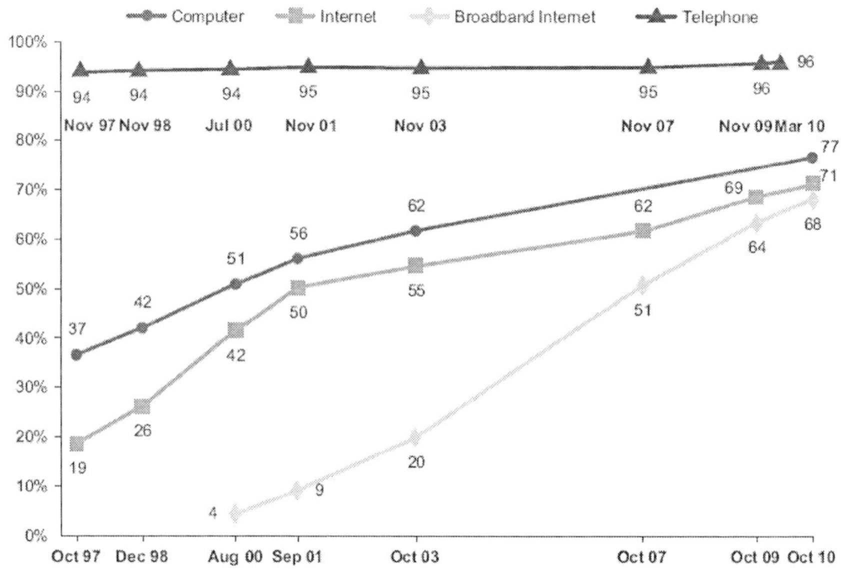

Source: U.S. Census Bureau, Current Population Survey School Enrollment and Internet Use Supplement, October 2010, and ESA calculations. "Digital Nation: Expanding Internet Usage," National Telecommunications and Information Administration, February 2011.

Note: Internet adoption depicted above combines broadband and dial-up Internet access services. 2001-2010 computer and Internet use data employ 2000 Census-based weights, and earlier years use 1990 Census-based weights.

Figure 1. Overview of Household Adoption Rates by Technology Percent of U.S. Households.

In February 2011, NTIA released *Digital Nation: Expanding Internet Usage*, a research preview that provided a first look at data from the October 2010 School Enrollment and Internet Use Supplement to the Census Bureau's Current Population Survey. The preview relied on several statistical tables that the Census Bureau released shortly after it began processing the survey data. These tables provided us with certain information about disparities in broadband adoption based on demographic characteristics such as income, race, and population density. The preliminary CPS data, however, limited our analysis until the Census Bureau made available a public use dataset for the survey, which facilitated a more detailed examination of the data.

Armed with complete survey results, ESA and NTIA produced custom statistics on adoption of Internet access services by demographics and geography, and reasons for non-adoption. In addition to calculating statistics not listed in the summary tables that NTIA used for the 2011 research preview, such as information on computer ownership, we employed regression analysis to estimate the effects of different variables, such as income or race, on such outcomes as broadband adoption when holding other factors constant. In other words, it allowed us to estimate how much of the digital divide remains attributable to population density or geography (for example) if we control for characteristics like education and income.

This report, therefore, draws on the October 2010 CPS data to provide new insights from the survey's expanded questions on computer use and Internet adoption.[2] These additional questions enabled us to study home computer use, identify the types of broadband Internet access services used at home, and determine the locations from which people used the Internet outside the home. In addition, one of the main findings of the 2011 preview was that affordability played a large role in a household's decision not to subscribe to broadband services. The expanded questionnaire allowed us to identify the types of costs that concerned households, including fixed costs, such as the cost of computer equipment and Internet service installation, and recurring costs, such as monthly Internet access subscription fees.

2. Household Computer Use, 2010

More than three-quarters (77 percent) of all American households had a computer at home in 2010, up from 62 percent in 2003.[3,4] Figure 2 shows that the majority of U.S. households (58 percent) used a desktop, laptop, netbook, or notebook computer only (personal computer), while 17 percent used a handheld device (which includes smartphones and other Internet-capable devices of similar functionality) in addition to a personal computer. Handheld devices appear to be complementary to personal computers since the substantial majority of households with handheld devices also used a personal computer. Only two percent of households reported having just a handheld device.

A significant segment of the population, almost one-fourth (23 percent) of all American households, did not own or use a computer at home in 2010.

The finding that a relatively small share of the population owned handheld devices raises the question whether households with handheld devices differ in their demographic characteristics from others. Table 1 shows demographic information for three groups: households that owned a handheld device (alone

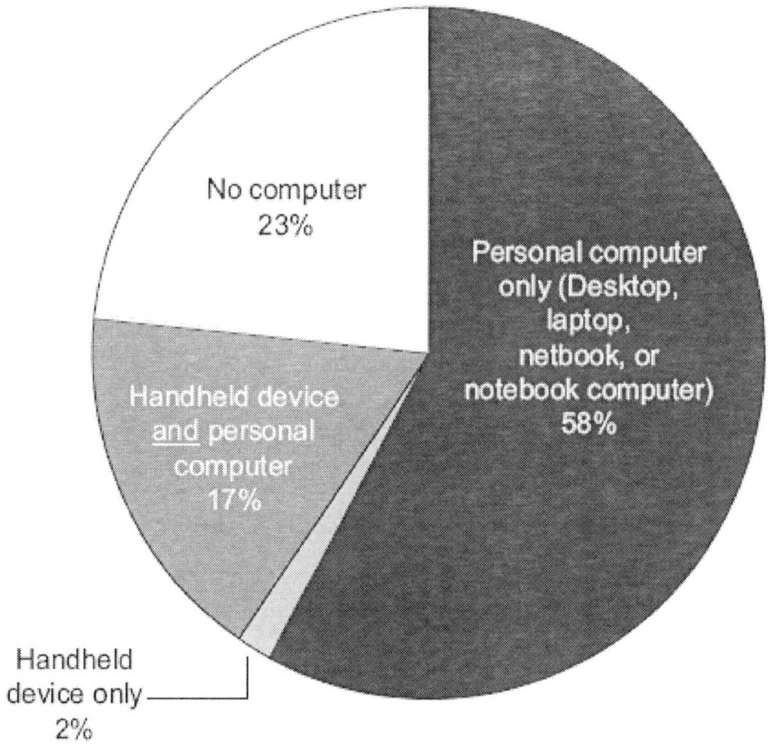

Source: U.S. Census Bureau, Current Population Survey School Enrollment and Internet Use Supplement, October 2010, and ESA calculations.

Note: Handheld devices include smartphones and other Internet-capable devices of similar size and functionality.

Figure 2. Household Computer Use by Type of Computer, 2010.

or in addition to a personal computer), those with a personal computer only, and those with no computer of any type.

Handheld devices were more prevalent in higher-income households – almost one-third of households (31 percent) with a handheld device had family incomes exceeding $100,000 in 2010, compared to 16 percent of households with a personal computer only, and three percent of households with no computer.

In general, households that reported using a handheld device had younger and more educated householders than those without a handheld device or computer.[5]

Table 1. Household Characteristics by Type of Computer at Home, 2010

	Handheld devicealone or incombination	Personal computeronly (Desktop, laptop, netbook, or notebook)	No computer
Mean Age* (years)	42	49	57
Income < $25,000	14%	23%	57%
Income > $100,000	31%	16%	3%
High school diploma*	18%	28%	42%
College degree or more*	48%	33%	10%
Urban	89%	83%	79%
Rural	10%	16%	21%

Source: U.S. Census Bureau, Current Population Survey School Enrollment and Internet Use Supplement, October 2010, and ESA calculations.

Note: Handheld devices include smartphones and other Internet-capable devices of similar functionality. *These are attributes of the householder.

3. Household Internet Adoption, 2010

3.1. Types of Household Internet Adoption

In 2010, more than two-thirds (68 percent) of all American households utilized broadband Internet access services, up four percentage points (64 percent) from the previous year.[6] Figure 3 shows that cable modem and DSL were, by far, the leading broadband technologies for residential Internet access, with 32 percent and 23 percent of households, respectively, utilizing these services.[7] Other broadband technologies, such as mobile broadband, fiber optics, and satellite technologies, accounted for a small segment of household Internet connections.

A shrinking share of home Internet users – about three percent of households in 2010 – used dial-up to access the Internet, down from five percent in 2009. By combining the percentages of households with broadband

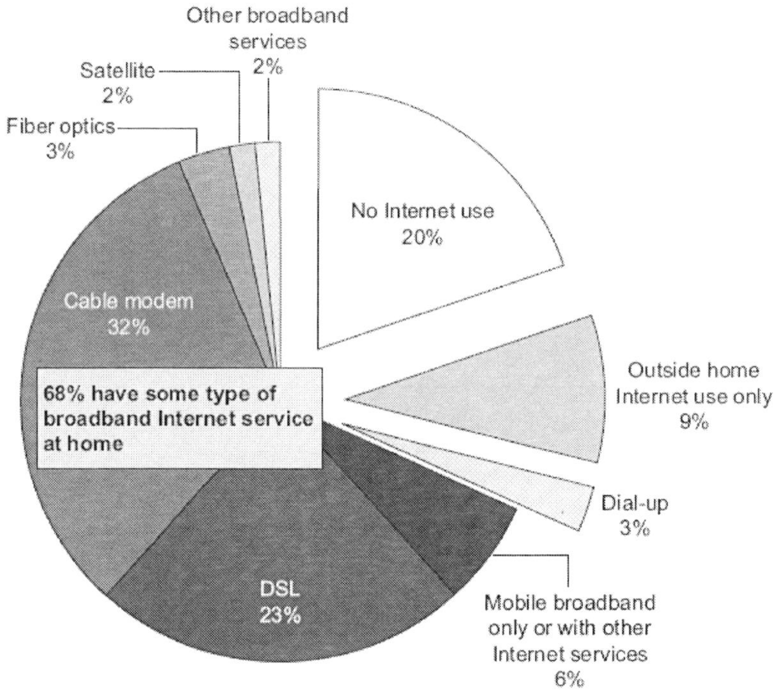

Source: U.S. Census Bureau, Current Population Survey School Enrollment and Internet Use Supplement, October 2010, and ESA calculations.

Note: Other broadband services include unspecified service types, or any other combination of listed service types excluding mobile broadband.

Figure 3. Household Internet Adoption by Type of Technology, 2010.

Internet access and those with dial-up services, we note that seven out of every ten (71 percent) American households had home Internet access service in 2010. Another nine percent of households had Internet users who only accessed the Internet outside the home. Together, these figures suggest that 80 percent of American households in 2010 had at least one Internet user, up three percentage points from the previous year.

3.2. Mobile Broadband Internet Use at Home

The explosive growth of wireless technologies in recent years reflects Americans' desire to carry portable devices that provide communications capabilities they previously could only access at home or work. Mobile broadband services offer Internet access utilizing service providers' cellular networks. Mobile broadband is unique in its ability to function wherever radio signals are available, rather than at a fixed location (or, where Wi-Fi routers are used, a small range of locations).

This feature supports a continuous Internet connection using mobile devices, and enables the use of location-aware online services. Smartphones with "data plans" represent the most common way people obtain mobile broadband service. In addition, mobile users may access mobile broadband services with cards, adapters, and base stations that connect computersand other Internet-ready devices.[8] CPS data suggest that a small share of households (six percent) utilized mobile broadband services at home in 2010 (Figure 3).[9]

Home mobile broadband adoption was more widespread in households reporting use of a handheld device. Figure 4 compares Internet use between two groups - households that reported owning a handheld device (alone or in addition to a personal computer) and households that did not own any handheld device but reported owning a personal computer. According to Figure 4, approximately one in five households (21 percent) with a handheld device used mobile broadband (Figure 4, right panel), compared to only four percent of households with personal computers (Figure 4, left panel).

In addition to having a stronger preference for mobile broadband Internet access, users of handheld devices were also more likely than their counterparts using only personal computers to subscribe to fiber optic broadband services and less likely to subscribe to DSL services. Six percent of households with a handheld device reported fiber optics use, compared to three percent of households with no handheld device but with a personal computer; DSL was used in 23 percent of households with a handheld device versus 33 percent of households with only a personal computer.

Figure 5 compares Internet use for urban and rural households. Note that this report uses metropolitan and nonmetropolitan areas as proxies for urban and rural areas, respectively.[10] Home mobile broadband penetration in 2010 was similar among urban (7 percent) and rural households (6 percent). Both urban and rural households report cable modem and DSL as the leading broadband technologies for residential broadband Internet access. Urban households, however, were more likely to use fiber optics and cable modem connections, and less likely to utilize DSL and dial-up services than their rural counterparts.

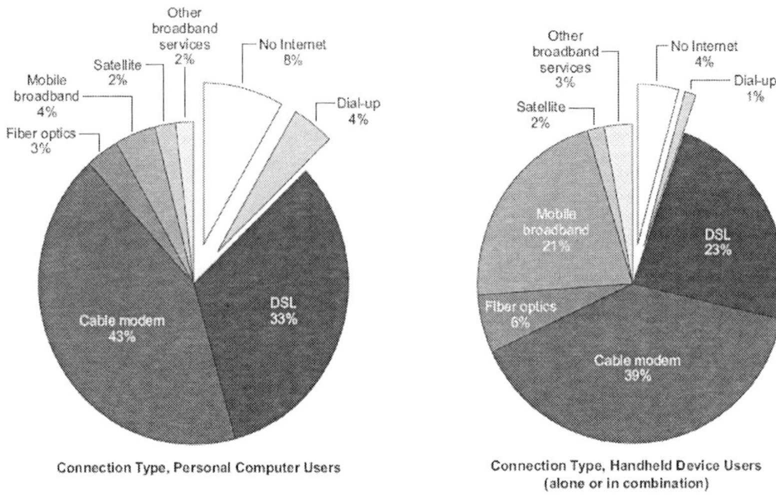

Source: U.S. Census Bureau, Current Population Survey School Enrollment and Internet Use Supplement, October 2010, and ESA calculations.

Figure 4. Internet Connection Technology by Home Computer Type, Households, 2010.

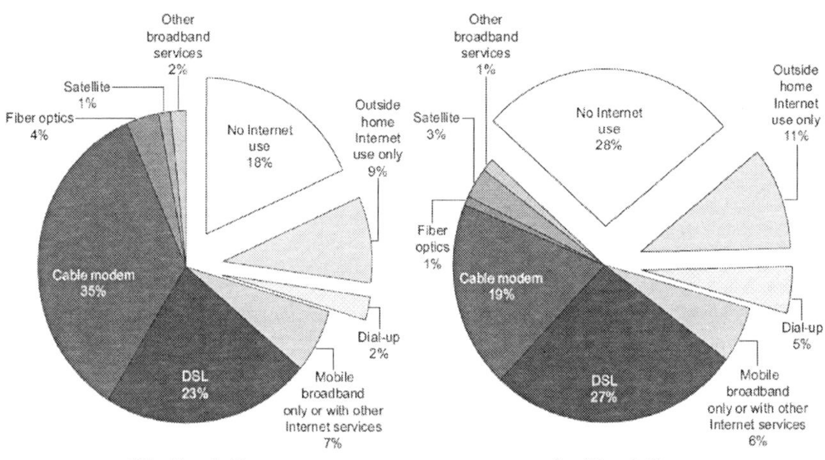

Source: U.S. Census Bureau, Current Population Survey School Enrollment and Internet Use Supplement, October 2010, and ESA calculations.

Figure 5. Internet Connection Type by Urban and Rural Location, 2010.

4. Computer and Home Internet Use by Household Demographic Characteristics and Geography, 2010

This section studies how computer use and home Internet adoption vary across households with different demographic and socio-economic backgrounds. It also examines the impact of geographic location on computer ownership and home Internet access. The results are useful to identify groups that lagged – or led – in adoption of these technologies in 2010. Sections 4.1 and 4.2 look at Internet adoption rates by demographic attributes and geography. Section 4.3 utilizes regression analysis to evaluate specifically broadband Internet adoption among comparable households.

The results indicate that households with lower incomes and less education, as well as Blacks, Hispanics, people with disabilities, and rural residents were less likely to have home Internet access service. However, differences in socio-economic attributes do not entirely explain why certain racial and ethnic groups or rural residents lagged in adoption. Further, households without computers comprised the vast majority of non-adopters of home broadband Internet access services. Predictably, the majority of computer users also utilized broadband at home.

4.1. Demographic and Geographic Gaps in Computer and Internet Use

Figure 6 shows that home computer use and Internet adoption are strongly associated with income. Almost half (46 percent) of the households in the lowest-income category did not have a computer, compared to only four percent of the highest-income households.

Focusing on broadband, adoption exhibited a similar relationship with income. Less than half (43 percent) of all households with annual household incomes below $25,000 in 2010 reported having broadband Internet access at home, compared with the vast majority (93 percent) of households with incomes exceeding $100,000. Dial-up service, however, accounted for a very small segment of households with Internet access irrespective of income (ranging from one percent to three percent of all households depending on income).

The total share of households with computers, as shown by the blue segments of each vertical bar in Figure 6, consists of households with broadband, dial-up, and those that reported having a computer, but no Internet access. The vast majority of this computer-using group had broadband Internet access at home. Moreover, this pattern was visible across income groups, suggesting that broadband adoption was more consistent among computer users than among all households across income groups. Almost four-fifths (79 percent) of households with computers and incomes below $25,000 used broadband at home, compared to 96 percent of computer-using households with incomes exceeding $100,000.[11]

Figure 7 illustrates home Internet access and computer ownership by presence of school-age children (between 6 and 17 years of age).

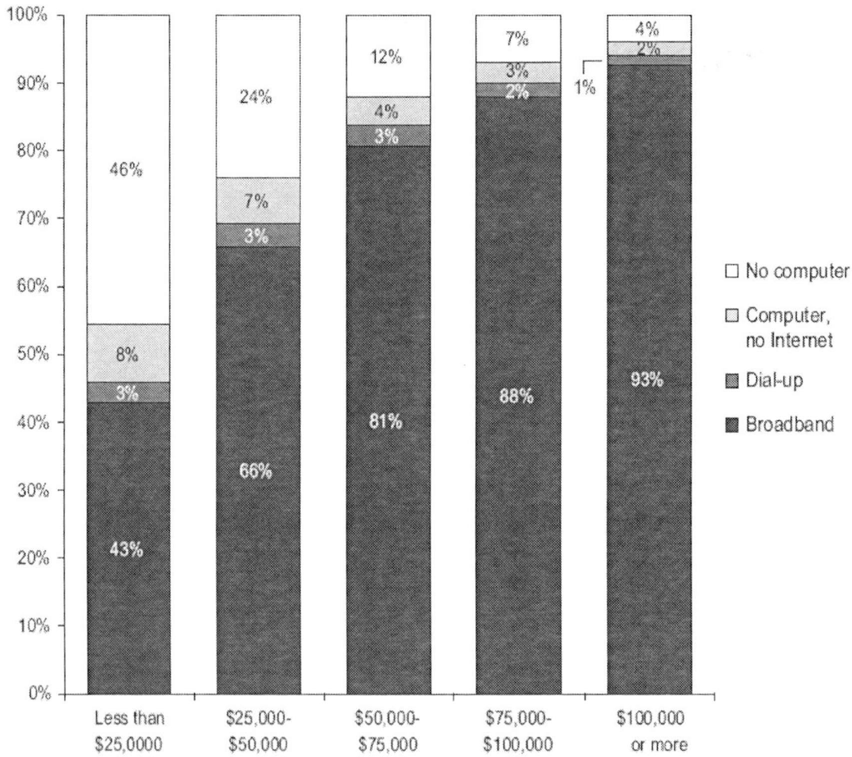

Source: U.S. Census Bureau, Current Population Survey School Enrollment and Internet Use Supplement, October 2010, and ESA calculations.

Figure 6. Computer and Internet Use by Household Income, 2010.

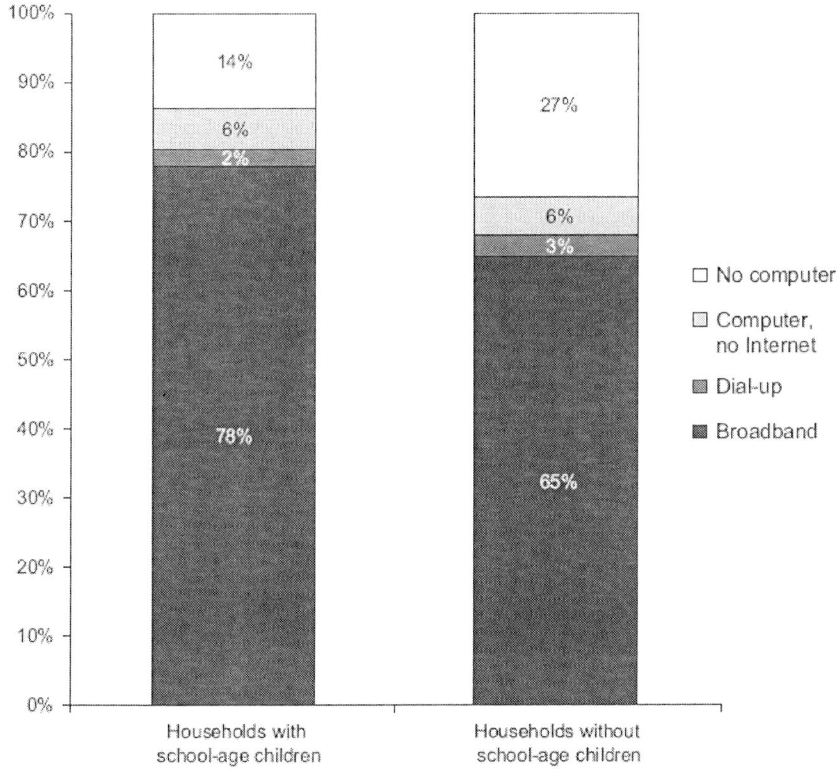

Source: U.S. Census Bureau, Current Population Survey School Enrollment and Internet Use Supplement, October 2010, and ESA calculations.

Figure 7. Computer and Internet Use by Presence of School-Age Children, 2010.

Households with one or more children between 6 and 17 years of age were more likely to own a computer (86 percent) and to have home broadband Internet access services (78 percent) than households with no school-age children (computer and broadband adoption rates were 74 percent and 65 percent, respectively).[12] Dial-up Internet access service was also less prevalent among households with school-age children (2 percent) than those without children (3 percent).

We find a similar relationship between home computer use and broadband adoption. The substantial majority of households with computers had broadband Internet access. Among computer owners, 91 percent of households with one or more school-age children, and 88 percent of households with no school-age child, utilized broadband (see Appendix B, Table B2).

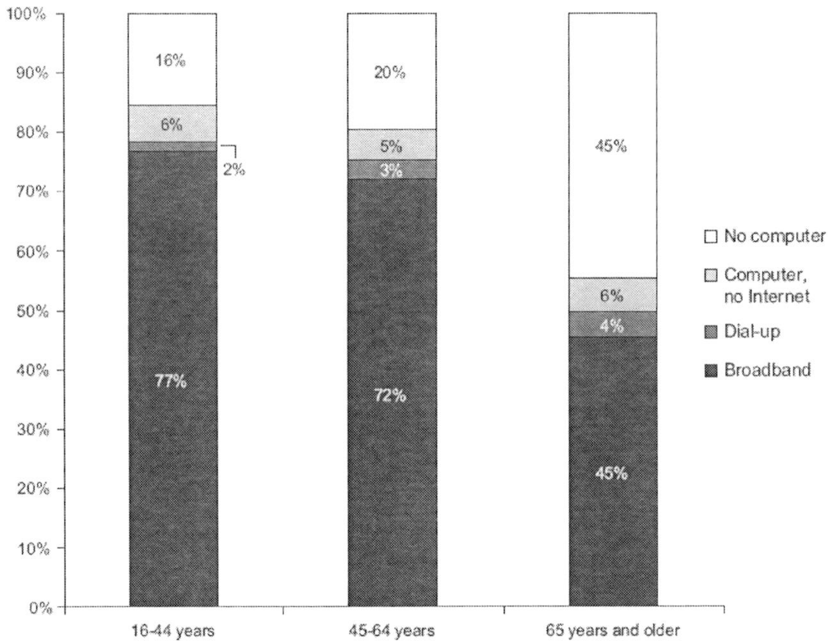

Source: U.S. Census Bureau, Current Population Survey School Enrollment and Internet Use Supplement, October 2010, and ESA calculations.

Figure 8. Computer and Internet Use by Householder Age, 2010.

Figure 8 shows home computer use and Internet adoption by householder age. Older householders, particularly those ages 65 and older, were less likely than their younger counterparts to live in a home with a computer (55 percent) or have broadband Internet access service at home (45 percent). Unsurprisingly, the differences in broadband adoption across age categories were much less pronounced among computer owners. Among households with computers, 91 percent of householders ages 16 to 44 years had broadband service, compared to 82 percent of their older counterparts, 65 years and older (see Appendix B, Table B2).

Figure 9 displays the data by race and ethnicity. Asian households exhibited the highest rates of home computer ownership (86 percent) and broadband service (81 percent), followed by White households (80 percent owned a computer and 72 percent had home broadband Internet services).[13] Hispanic households and Black households lagged behind – only about two-thirds of Black households and Hispanic households (65 percent and 67 percent, respectively) had a computer at home, and only slightly more than

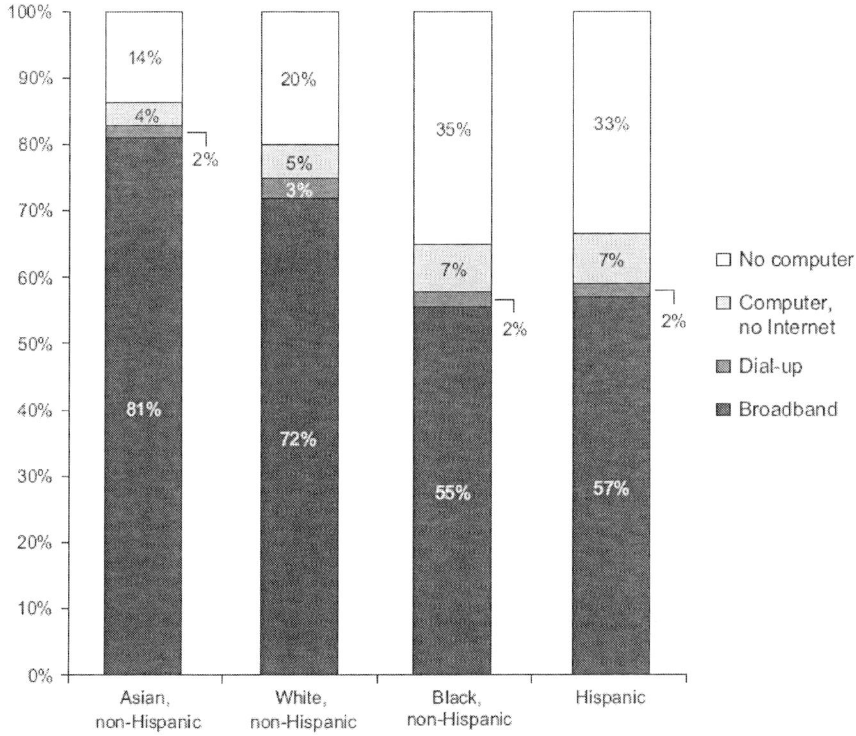

Source: U.S. Census Bureau, Current Population Survey School Enrollment and Internet Use Supplement, October 2010, and ESA calculations.

Note: Data for Whites, Blacks, Asians, and American Indians and Alaska Natives do not include people of Hispanic origin. Persons of Hispanic origin may be of any race.

Figure 9. Computer and Internet Use by Householder Race and Ethnicity, 2010.

half of all Black and Hispanic households (55 percent and 57 percent, respectively) had broadband service.[14] Households headed by AmericanIndian or Alaska Native householders also had computer use (66 percent) and broadband adoption (52 percent) rates that trailed the national average (see Appendix B, Table B2).[15]

Data for only computer-using households show a less pronounced race- and ethnicity-related gap in broadband adoption. Among households using computers, broadband Internet adoption rates were 94 percent for Asians, 90 percent for Whites, and 86 percent for both Black and Hispanic households (see Appendix B, Table B2). This again suggests that computer use is strongly correlated with broadband Internet access at home.

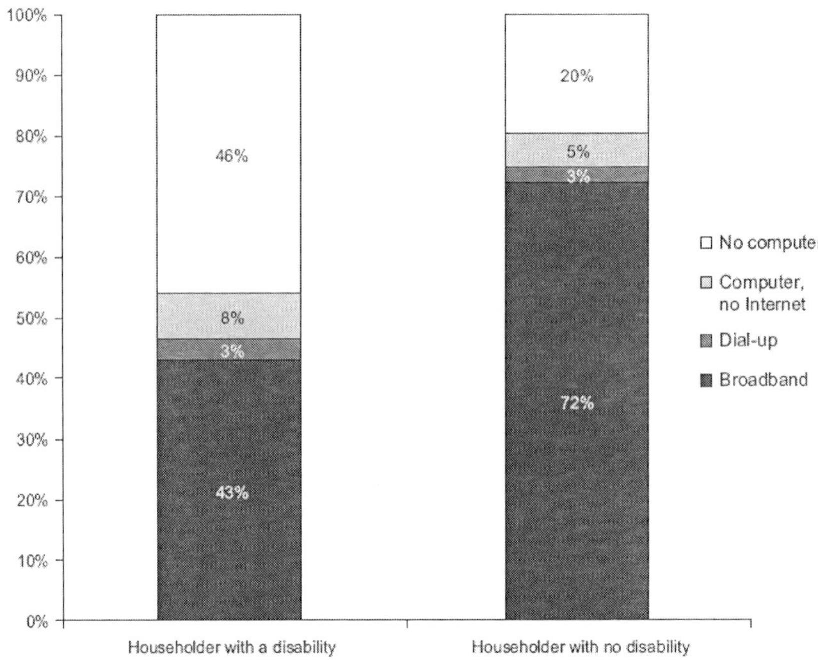

Source: U.S. Census Bureau, Current Population Survey School Enrollment and Internet Use Supplement, October 2010, and ESA calculations.

Figure 10. Computer and Internet Use by Householder Disability Status, 2010.

Figure 10 shows home computer use and Internet adoption by householder's disability status.[16] Almost half of all households headed by someone with a disability did not have a computer at home (46 percent), compared to a much smaller segment (20 percent) of homes where the householder had no disability. Households headed by people with disabilities were also much less likely to subscribe to broadband service than those with no disability.

Among computer-using households, 80 percent of homes headed by someone with a disability used broadband at home, compared to 90 percent of those headed by someone with no disability (see Appendix B, Table B2).

The next three figures show household computer ownership and Internet use data by geographic location. Figure 11 demonstrates that one out of five urban households (22 percent) did not own a computer in 2010, compared to roughly one out of three (30 percent) rural households. Urban (metropolitan) dwellers also were more likely than their rural (non-metropolitan) counterparts to have broadband Internet access at home (70 percent compared to 57 percent). Rural

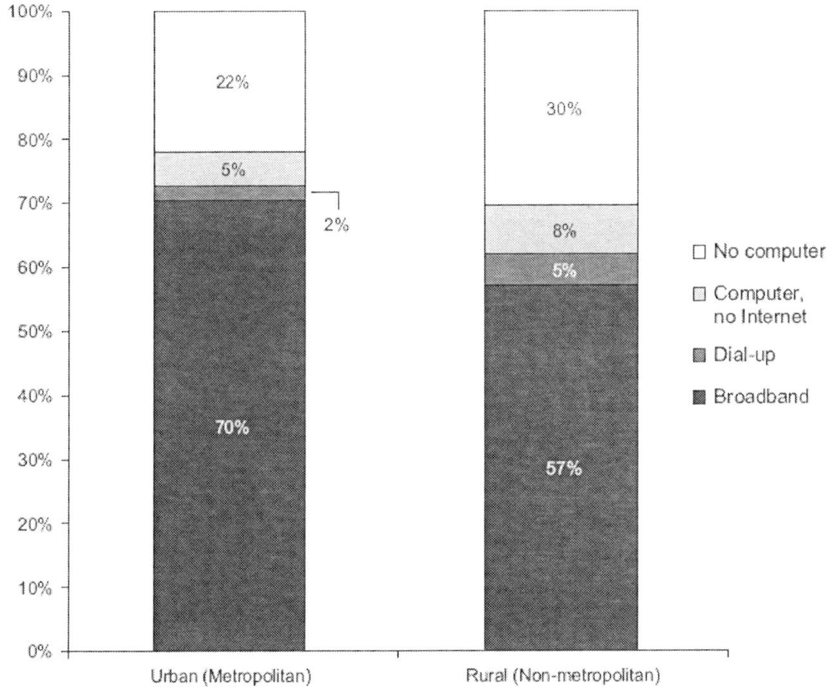

Source: U.S. Census Bureau, Current Population Survey School Enrollment and Internet Use Supplement, October 2010, and ESA calculations.

Figure 11. Computer and Internet Use by Urban and Rural Location, 2010.

residents were more likely to utilize dial-up services to go online – five percent of rural households utilized dial-up services, compared with two percent of urban dwellers.

Among computer owners, 90 percent of urban households and 82 percent of rural households used broadband services (see Appendix B, Table B2).

Figure 12 shows computer and Internet use by state. Proportions of households without a computer at home ranged from about one-third of households (33 percent) in Mississippi to 13 percent of households in Utah. Average broadband adoption in 2010 varied by state from about half (52 percent) of all households in Mississippi to 80 percent in Utah. Broadband Internet access service was the leading mode for residential Internet access, while dial-up service accounted for a small segment of online households regardless of state.[17]

The pattern of generally higher broadband adoption in urban compared to rural areas occurred across states. Figure 13 shows the share of urban and rural households with broadband service by state. The urban-rural gap in broadband adoption varied from zero percent in California (that is, average broadband adoption rates were practically the same in urban and rural areas of California) to 26 percent in Mississippi, where the average broadband adoption rates ranged from 67 percent in urban areas to 41 percent in rural areas.[18]

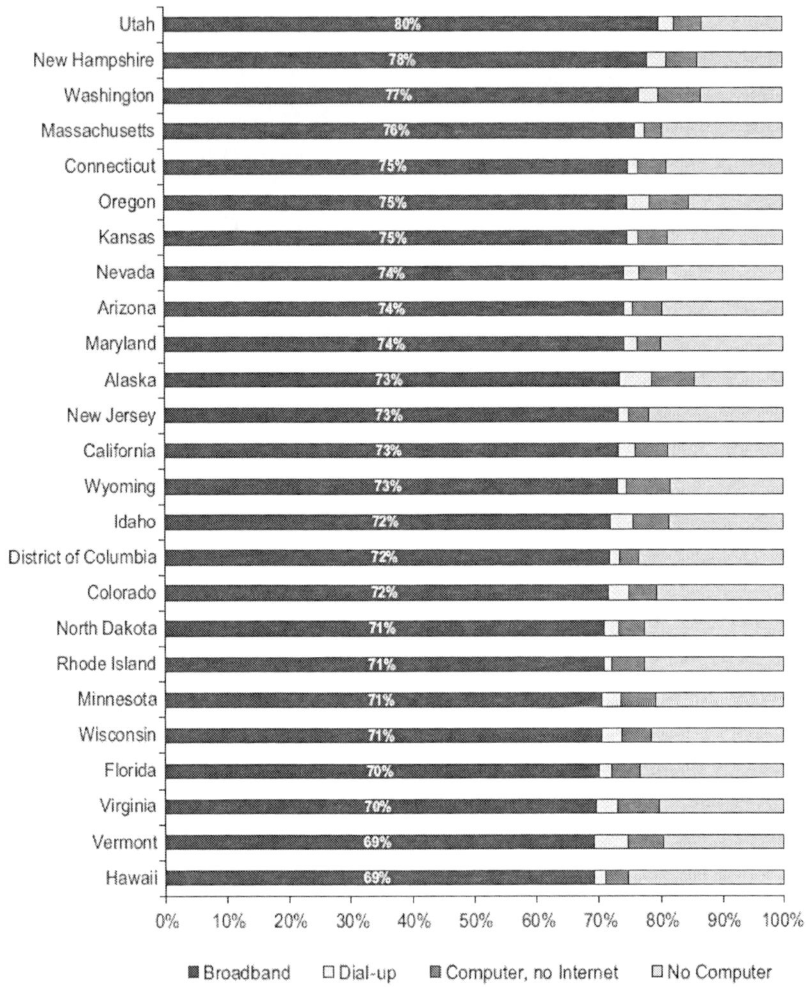

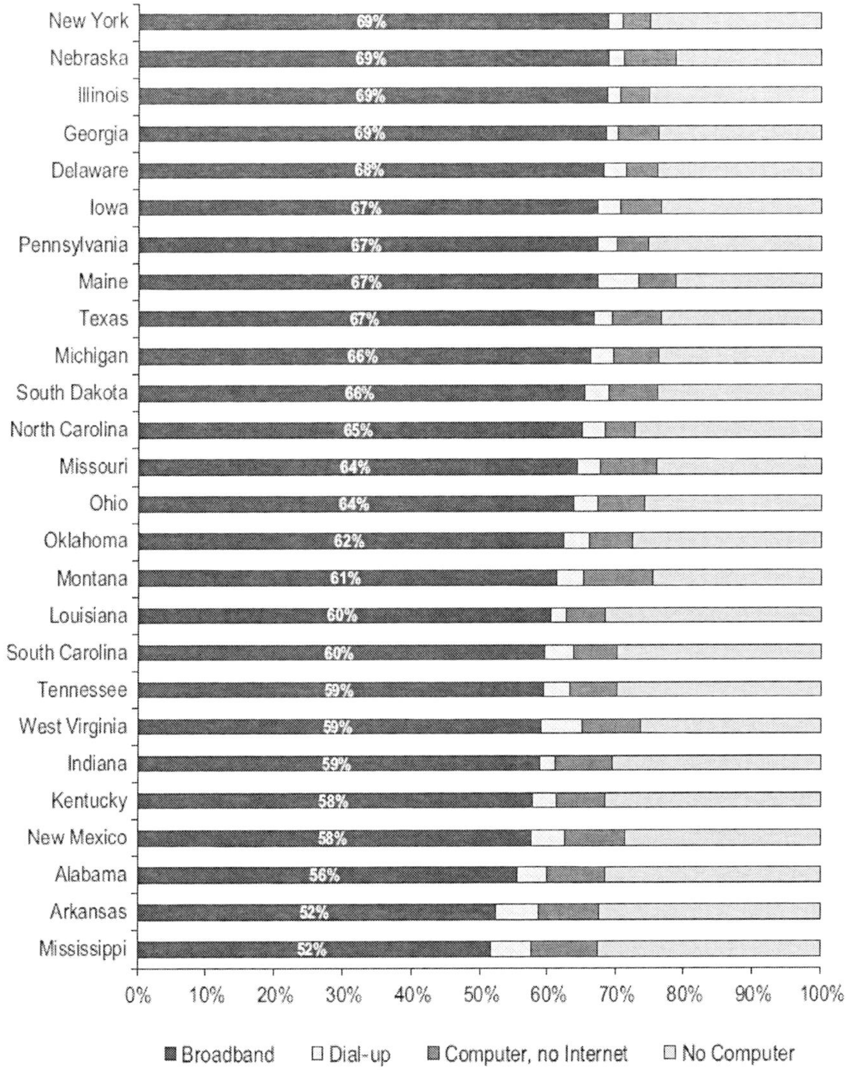

Source: U.S. Census Bureau, Current Population Survey School Enrollment and Internet Use Supplement, October 2010, and ESA calculations.

Note: Because of sampling variability, average adoption rates for two states may not be different from one another in a statistically significant way. Tables B5 and B6 in Appendix B provide the 90 percent confidence interval for each state.

Figure 12. Computer and Internet Use by State, 2010 (by Household Broadband Adoption Rate).

4.2. A Detailed Look at Computer and Broadband Internet Use by Demographic Characteristics and Geography

This section provides a more detailed analysis of the survey results with respect to demographic characteristics and geography. The previous section identified the households exhibiting low and high levels of computer use and Internet access service adoption, and separately, their urban or rural locations. This section combines the "who" and "where" in our identification of low and high adopters.

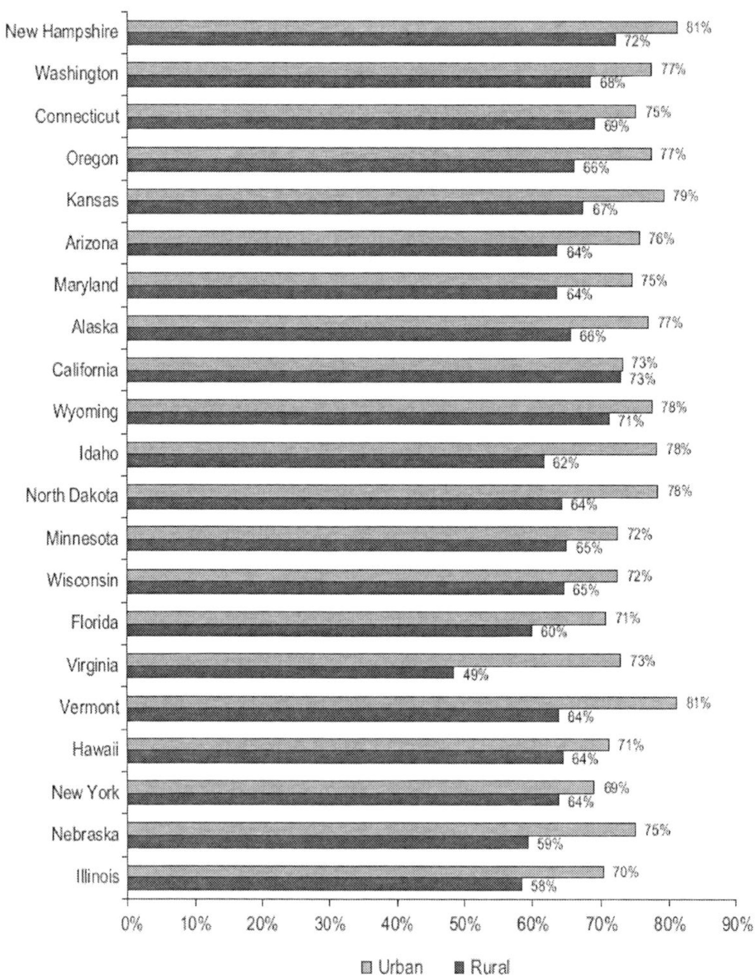

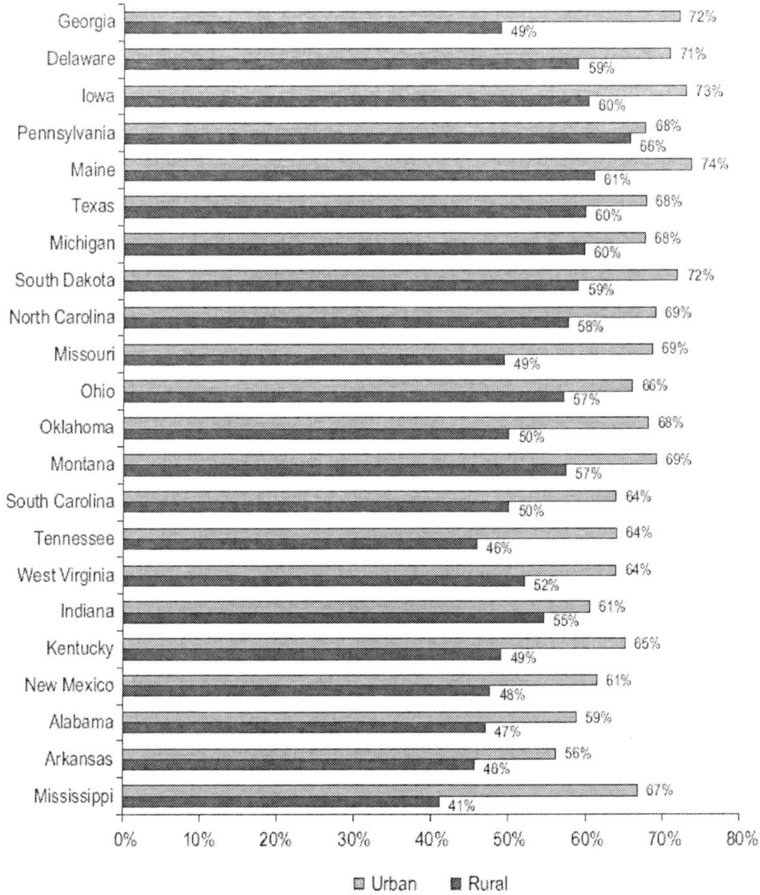

Source: U.S. Census Bureau, Current Population Survey School Enrollment and Internet Use Supplement, October 2010, and ESA calculations.

Note: The following states were not included in this figure due to data limitations for rural areas in the CPS: Colorado, Louisiana, Nevada, Rhode Island, Massachusetts, New Jersey, and Utah. The District of Columbia had no rural areas. In 2010, the average broadband adoption rates for the urban areas in these states were as follows: Colorado - 71 percent, Louisiana - 63 percent, Nevada - 75 percent, Rhode Island - 71 percent, Massachusetts - 76 percent, New Jersey – 73 percent, and Utah - 79 percent. Because of sampling variability, average adoption rates for two states may not be different from one another in a statistically significant way. Tables B7 and B8 in Appendix B provide the 90 percent confidence interval for each state.

Figure 13. Broadband Internet Use in Urban and Rural Locations, by State, 2010 (by Overall Household Broadband Adoption Rate).

Table 2. Household Computer Use and Broadband Internet Adoption by Urban/Rural Location, Race, Ethnicity, Income, and Education, 2010

Household Characteristic	Computer Use		Broadbannd Adoption	
	Urban	Rural	Urban	Rural
All Households	78%	70%	70%	57%
Race and Ethnicity*				
White, non-Hispanic	82%	72%	75%	60%
Black, non-Hispanic	66%	53%	57%	41%
Hispanic	67%	57%	58%	46%
Asian, non-Hispanic	86%	85%	81%	83%
American Indian and Alaska Native, non-Hispanic	74%	52%	66%	31%
Household Income				
Less than $25,0000	56%	49%	45%	35%
$25,000-$50,000	76%	74%	67%	60%
$50,000-$75,000	88%	87%	82%	76%
$75,000-$100,000	93%	91%	89%	82%
$100,000 or more	96%	94%	93%	87%
Education*				
No high school diploma	46%	39%	35%	26%
High school diploma	68%	64%	59%	50%
Some college	84%	82%	75%	69%
College degree or more	93%	89%	88%	80%

Source: U.S. Census Bureau, Current Population Survey School Enrollment and Internet Use Supplement, October 2010, and ESA calculations.

Note: Data for Whites, Blacks, Asians, and American Indians and Alaska Natives do not include people of Hispanic origin. Persons of Hispanic origin may be of any race.

*These are attributes of the householder.

Table 2 shows computer use and broadband Internet adoption by race, ethnicity, income, and education within urban and rural areas. Households headed by someone without a high school diploma had particularly low home computer use and broadband service adoption, while rural households displayed slightly lower rates than their urban counterparts. Among households headed by someone without a high school diploma, 39 percent of rural households and 46 percent of urban households used a computer; and 26 percent of rural households and 35 percent of urban households had broadband service. Compared to the national average of 68 percent, broadband Internet adoption was also particularly low among rural households with incomes less than $25,000 (35 percent), and rural households headed by someone of American Indian or Alaska Native background (31 percent).

Table 3 contains data on computer use and broadband Internet adoption by race, ethnicity, family income, and urban or rural location. Rural Black and Hispanic households with family incomes below $25,000 had the lowest rates of home computer use (44 percent for Blacks, 45 percent for Hispanics) and broadband Internet adoption (32 percent for Blacks, 30 percent for Hispanics).

Table 3. Household Computer Use and Broadband Internet Adoption by Urban/Rural Location, Race, Ethnicity and Income, 2010

Race, Ethnicity, and Income	Computer Use		Broadbannd Adoption	
	Urban	Rural	Urban	Rural
All Households	78%	70%	70%	57%
White, non-Hispanic				
Less than $25,0000	60%	50%	49%	36%
$25,000-$50,000	78%	75%	69%	62%
$50,000-$75,000	89%	87%	83%	76%
$75,000 or more	96%	93%	92%	85%
Black, non-Hispanic				
Less than $25,0000	49%	44%	39%	32%
$25,000-$50,000	70%	64%	60%	47%
$50,000-$75,000	83%	77%	76%	70%
$75,000 or more	90%	85%	84%	81%
Hispanic				
Less than $25,0000	49%	45%	38%	30%
$25,000-$50,000	71%	60%	60%	53%
$50,000-$75,000	84%	85%	77%	79%
$75,000 or more	93%	86%	88%	76%

Source: U.S. Census Bureau, Current Population Survey School Enrollment and Internet Use Supplement, October 2010, and ESA calculations.

Note: Data for Whites, Blacks, Asians, and American Indians and Alaska Natives do not include people of Hispanic origin. Persons of Hispanic origin may be of any race. Non-Hispanic Asians, and non-Hispanic American Indians and Alaska Natives are not included in this table due to small sample issues.

Table 4 displays a similar pattern of computer use and broadband Internet adoption by race and educational attainment, and by urban or rural location. Households headed by a Black householder without a high school diploma and living in rural areas exhibited the lowest levels of home computer use (27 percent) and broadband Internet adoption (16 percent).

Table 4 Household Computer Use and Broadband Internet Adoption by Urban/Rural Location, Race, Ethnicity and Education 2010

Householder Characteristics	Computer Use		Broadbannd Adoption	
All Households	Urban	Rural	Urban	Rural
	78%	70%	70%	57%
White, non-Hispanic				
No high school diploma	47%	42%	37%	27%
High school diploma	70%	65%	62%	52%
Some college	86%	83%	78%	70%
College degree or more	93%	90%	89%	80%
Black, non-Hispanic				
No high school diploma	36%	27%	27%	16%
High school diploma	57%	50%	47%	38%
Some college	74%	73%	64%	56%
College degree or more	87%	82%	81%	75%
Hispanic				
No high school diploma	47%	38%	35%	28%
High school diploma	65%	59%	56%	45%
Some college	83%	81%	74%	73%
College degree or more	90%	83%	84%	72%

Source: U.S. Census Bureau, Current Population Survey School Enrollment and Internet Use Supplement, October 2010, and ESA calculations.

Note: Data for Whites, Blacks, Asians, and American Indians and Alaska Natives do not include people of Hispanic origin. Persons of Hispanic origin may be of any race. Non-Hispanic Asians, and non-Hispanic American Indians and Alaska Natives are not included in this table due to small sample issues.

4.3. Marginal Effects of Household Characteristics on the Likelihood that a Household Uses Broadband Internet Access Service

The finding that socio-economic characteristics, as well as race, ethnicity, and geographic location are highly correlated with household technology usage patterns might be misleading, as these household attributes are themselves correlated with each other. For instance, income and education are likely to be higher in urban areas if employment opportunities requiring high levels of skill and specialization are disproportionately located in urban areas. As a result, it is not clear from the tabulations how much of the urban-rural gap in home broadband adoption results from socio-economic differences between urban and rural residents. The same issue applies for the tabulations by race and ethnicity.

By utilizing regression analysis, we can estimate the marginal or "isolated" association between technology usage and a particular household characteristic.[19]

For example, one way of estimating the marginal effect of living in an urban area on broadband adoption is by comparing broadband adoption among urban and rural households that are otherwise similar with respect to key determinants of broadband adoption, such as income, education, race, ethnicity, age, state of residence, and other factors.

In other words, the marginal effect of living in an urban rather than a rural location is the gap in average broadband adoption between urban and rural households, after accounting for differences in socio-economic and demographic characteristics.

The rest of this section utilizes regression analysis to estimate the relationship between selected demographic and geographic characteristics and home broadband Internet adoption.

The factors for which we control in this analysis include household income, education, age, race, ethnicity, foreign-born status, household composition (total number of persons in a household and whether a related school-age child lives there), disability status, and geographic location (urban-rural location and state). For characteristics like education, race, ethnicity, age, disability status, and foreign-born status, we use information for the householder.

Note that the CPS data do not provide information on availability and price of Internet access services in a household's immediate location, both of which are important determinants of adoption.

Even though we are unable to control directly for price and availability, the regression analysis accounts for a household's geographic location (urban or rural location, the population size of a household's urban area, and state) and therefore would capture some of the variation in price and availability along these geographic dimensions.[20]

We present the results from this regression analysis in Table B4 of Appendix B. Figures 14-18 graphically present the gaps in broadband adoption.

Each figure uses a pair of bars to display the adoption gap between two groups of households. Within each pair, the left bar (which is also the longer bar) shows the simple gap in average adoption before controlling for other household attributes. The right bar (which is also the shorter bar) shows the remaining adoption gap which is unexplained by our model, that is, it is the remaining gap after accounting for differences in household demographic, socio-economic, and geographic characteristics. Within each figure, the left panel (or pair of bars) presents the adoption gap information based on data for all households, and the right panel shows the information based on the sample of computer owners only.

Figure 14 shows the adoption gaps between urban and rural households. According to Figure 11 from the previous section, 70 percent and 57 percent of urban and rural households, respectively, reported having broadband service at home, identifying a 13 percentage point gap in broadband adoption between urban and rural households (represented by the left bar in the left panel of Figure 14).

After we account for socio-economic and demographic differences between urban and rural households, the remaining adoption gap is five percentage points (represented by the right bar in the left panel of Figure 14).

The right panel of Figure 14 shows the urban-rural adoption gap among computer users only. Ninety percent of urban households with computers and 82 percent of rural households with computers reported having broadband at home, indicating an eight percentage point gap. After accounting for socio-economic and demographic attributes, the gap declines to five percentage points.

This suggests that, even among computer owners of similar income, education, age, and other demographic characteristics, urban dwellers were on average five percentage points more likely than their rural counterparts to adopt home broadband service.

The variation in price and availability of broadband services could account for part of this unexplained urban-rural gap.

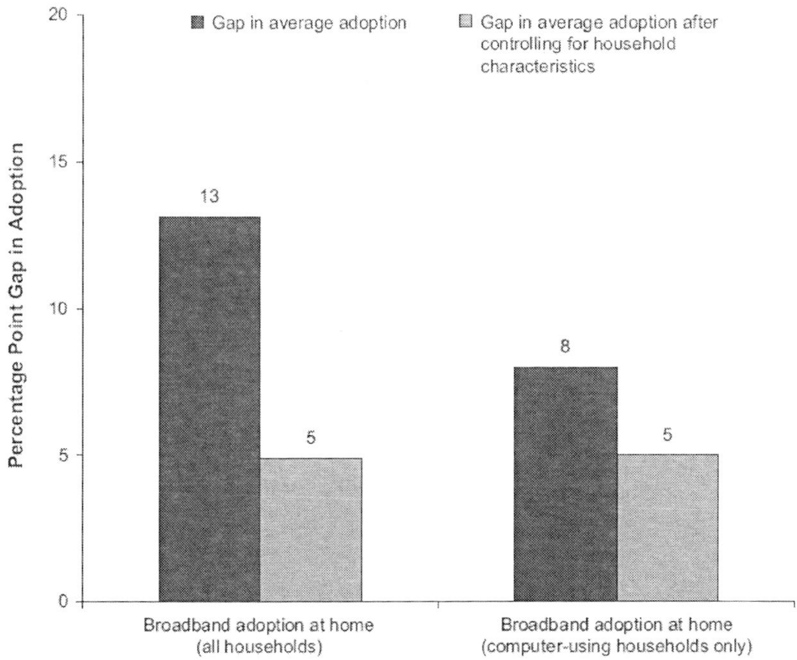

Source: U.S. Census Bureau, Current Population Survey School Enrollment and Internet Use Supplement, October 2010, and ESA calculations.

Figure 14. Broadband Internet Adoption Gap between Urban and Rural Households, 2010.

Turning to the data by race and ethnicity, Figure 9 showed that Asian households displayed the highest rates of broadband adoption (81 percent), followed by White (72 percent), Hispanic (57 percent), and Black (55 percent) households. This suggests that Asian households on average were nine percentage points more likely to have broadband Internet access services than White households. In addition, White households were 17 percentage points and 15 percentage points more likely than Black and Hispanic households, respectively, to have home broadband Internet access. Once we control for socio-economic and geographic differences, the broadband adoption gap between Asians and Whites disappears (Figure 15), whereas the gap between White and Black households, and between White and Hispanic households both decline to 11 percentage points (left panels, Figures 16 and 17). An important topic for future research would be the persistence of broadband adoption gaps between the latter groups even after accounting for demographic, socioeconomic, and geographic factors.

Computer users, on the contrary, displayed much less disparate broadband adoption rates across race and ethnicity. Ninety-four percent of Asian households, 90 percent of White households, 86 percent of Black households, and 86 percent of Hispanic households with computers used broadband at home, implying adoption gaps of four percentage points between Asian and White, White and Black, and White and Hispanic households. Controlling for demographic characteristics and geography erases the Asian and White difference, and reduces both the White-Black and White-Hispanic gaps to three percentage points.

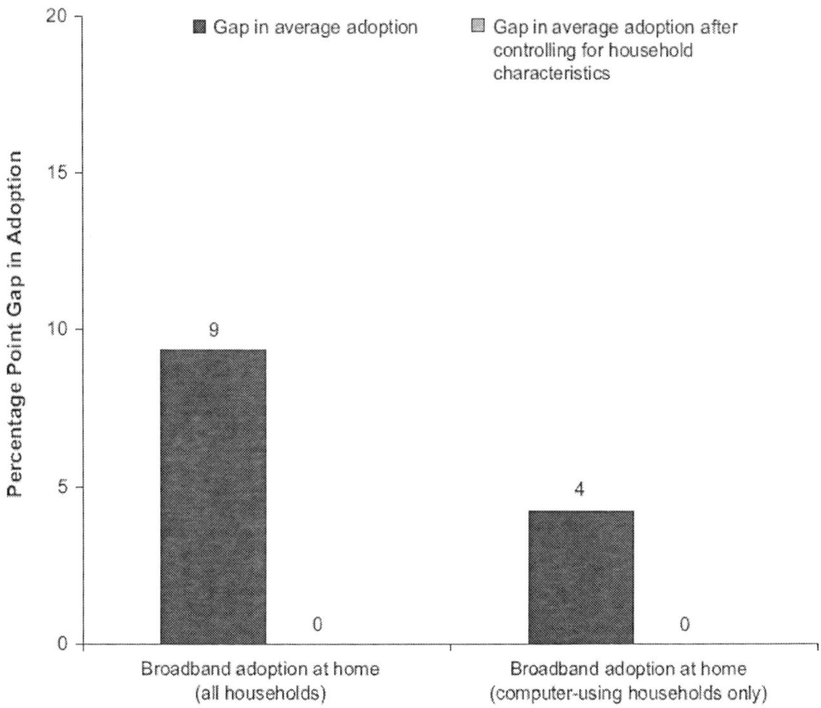

Source: U.S. Census Bureau, Current Population Survey School Enrollment and Internet Use Supplement, October 2010, and ESA calculations.

Note: Data for Whites, Blacks, Asians, and American Indians and Alaska Natives do not include people of Hispanic origin. Persons of Hispanic origin may be of any race.

Figure 15. Broadband Internet Adoption Gap between non-Hispanic Asian and White Households, 2010.

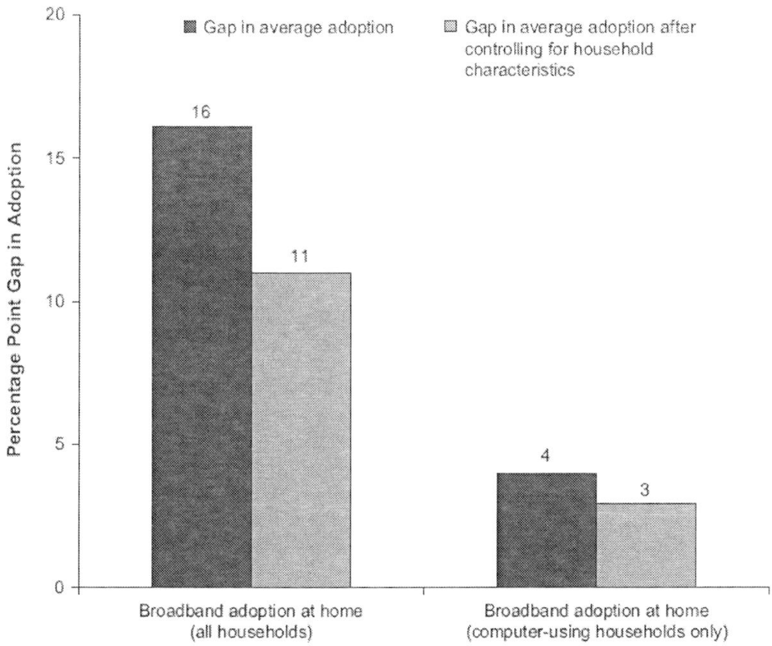

Source: U.S. Census Bureau, Current Population Survey School Enrollment and Internet Use Supplement, October 2010, and ESA calculations.

Note: Data for Whites, Blacks, Asians, and American Indians and Alaska Natives do not include people of Hispanic origin. Persons of Hispanic origin may be of any race.

Figure 16. Broadband Internet Adoption Gap between non-Hispanic White and Black Households, 2010.

Turning to the data by disability status, Figure 10 showed 43 percent of households headed by someone with a disability used broadband, compared to 72 percent of households headed by someone with no disability, implying a 29 percentage point gap in broadband Internet access. Once we control for income, education, age, and other key attributes, the gap in broadband Internet access declines to six percentage points or about one-fifth of the original gap (see left panel, Figure 18). Looking at computer owners exclusively shows smaller differences (10 percentage point difference before and three percentage point difference after controlling for demographic factors and geography, as shown in Figure 18, right panel). This suggests that differences in demographic and socio-economic attributes and geography explain a substantial portion of the disability-related broadband gap, even among computer owners.

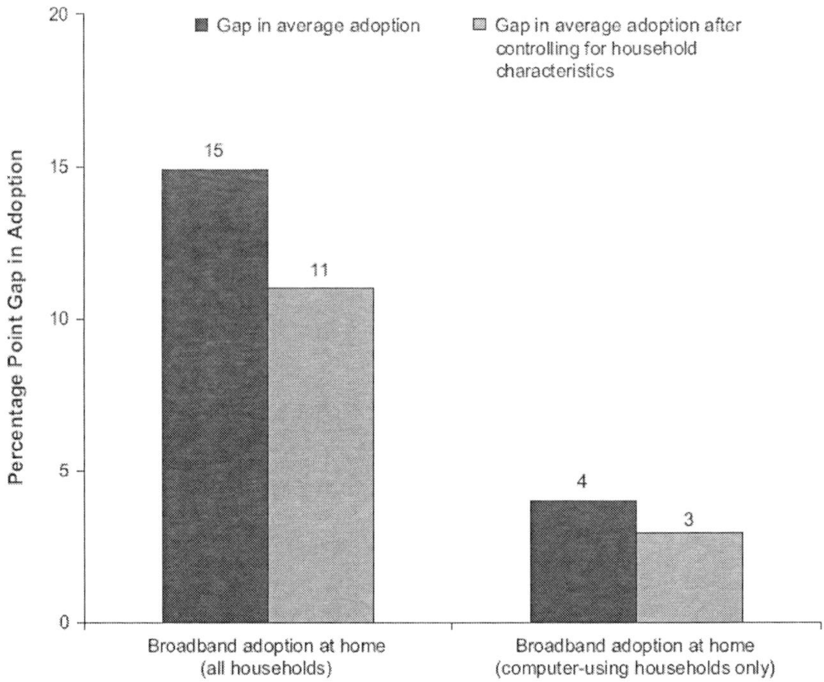

Source: U.S. Census Bureau, Current Population Survey School Enrollment and Internet Use Supplement, October 2010, and ESA calculations.

Note: Data for Whites, Blacks, Asians, and American Indians and Alaska Natives do not include people of Hispanic origin. Persons of Hispanic origin may be of any race.

Figure 17. Broadband Internet Adoption Gap between non-Hispanic White and Hispanic Households, 2010.

The regression results presented in this section suggest that income and education are strongly associated with broadband Internet access at home, but are not the sole determinants. Other factors, such as race, ethnicity, and urban-rural location are also independently associated with technology usage patterns. The adoption gaps are less pronounced among computer users, that is, computer use and broadband adoption strongly correlate, perhaps because both exhibit strong associations with demographic and socio-economic attributes.

The CPS data do not provide information on price and availability of broadband Internet access in a household's immediate location, which is why we are unable to directly account for these factors. Consequently, we are unable to distinguish how much of the variation across socioeconomic and

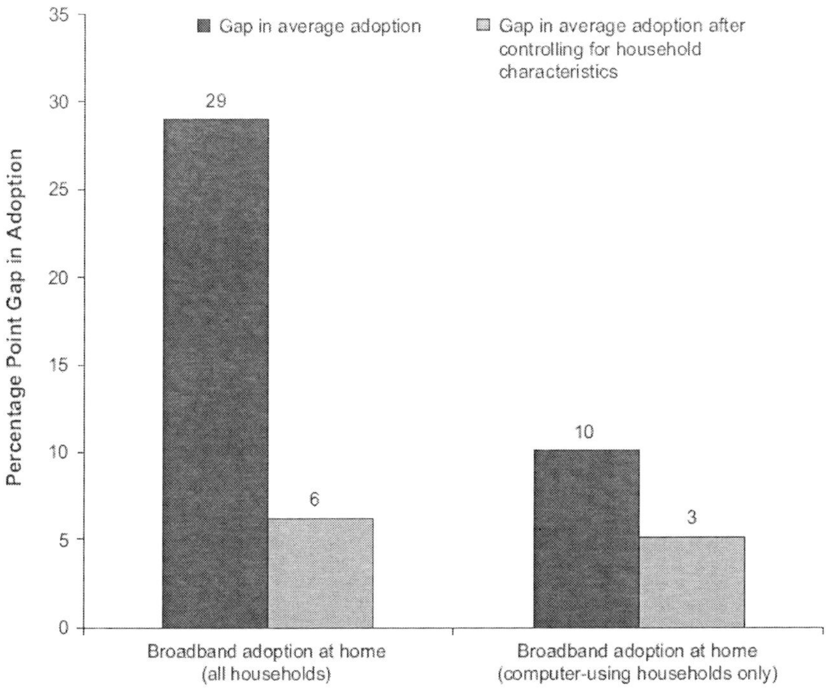

Source: U.S. Census Bureau, Current Population Survey School Enrollment and Internet Use Supplement, October 2010, and ESA calculations.

Figure 18. Broadband Internet Adoption Gap between Householders with no Disability and with Disability, 2010.

geographic dimensions results from factors related to demand as opposed to supply considerations. Lower demand for broadband Internet access or lack of affordability may partially explain why some households decline to adopt broadband Internet access service, but other reasons may include lack of supply or availability of residential broadband services.\

5. Non-Adoption of Internet Access Services at Home, 2010

The CPS Supplement asked households using dial-up Internet access services (hereafter referred to as "dial-up households") to state their main reason for not having broadband Internet service at home (tabulations shown on the left panel of Figure 19).

In addition, the CPS asked households without Internet service or a home computer to state their main reason for not having home Internet access (tabulations shown on the right panel of Figure 19).

The reasons for non-adoption are likely to have significant policy implications. For example, if lack of availability of broadband services is the main impediment to broadband adoption at home, then policies to expand usage may require attracting broadband providers to offer service. However, if there is a lack of information about broadband service availability, or a perceived lack of need or interest in broadband, then policies may incorporate public awareness campaigns.

Note that households' responses on the main reason for not using broadband Internet at home reflect their subjective opinions since the survey respondent may not have adequate information on pricing, availability, or the value of Internet access services. For instance, one may believe that broadband Internet access is not available in the area, but be misinformed. As a result, we caution that any comparison across households, while informative, requires careful analysis.

The left panel of Figure 19 shows that expense, lack of need, and lack of availability were the main impediments to broadband adoption for dial-up households.

One-third of dial-up users reported "lack of need" (33 percent), about one-third reported "too expensive" (34 percent), and about one-fourth (27 percent) reported "lack of availability."

Note that dial-up households accounted for three percent of American households in 2010.

In contrast, almost half (47 percent) of households without a computer or home Internet access stated lack of need as their main reason for not having home Internet services (right panel of Figure 19). About one-fourth (24 percent) reported affordability, and 15 percent reported inadequate computer as the primary reason for no home Internet access. Note that those households reporting no Internet access or home computer were a much larger group (29 percent) than the collective dial-up households (representing about three percent of American households).

Figure 20 tabulates the main reason for no Internet access at home separately for those households with, and those without, a computer at home. Among households that owned a computer but did not have home Internet access, expense was the most commonly provided reason (37 percent), followed by lack of need (28 percent). On the contrary, among households that did not own a computer, lack of need or demand dominated (52 percent).

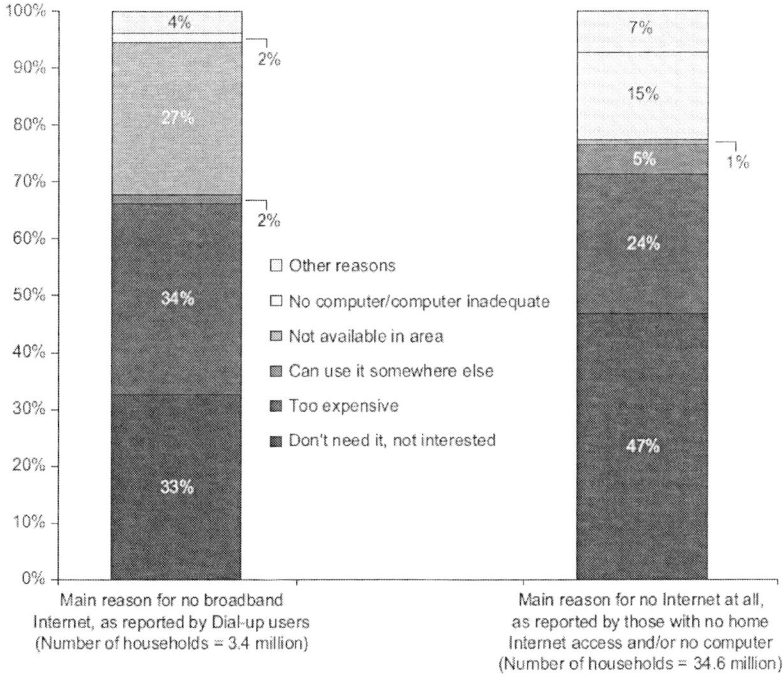

Source: U.S. Census Bureau, Current Population Survey School Enrollment and Internet Use Supplement, October 2010, and ESA calculations.

Figure 19. Main Reason for Not Having Home Internet Access Service, 2010.

The CPS asked households that responded "too expensive" as their main reason for not having home Internet service to indicate what costs most concerned them – the fixed costs of purchasing a computer or installing home Internet access services, or the recurring monthly subscription costs. Figure 21 tabulates these responses. The left panel of Figure 21 shows that, among dial-up households stating expense or affordability as their main reason for not having broadband, the cost of monthly Internet access service was a more serious concern than fixed costs. The vast majority (75 percent) of these households cited the monthly service cost and another 10 percent reported both the monthly service cost and fixed costs as their main impediments to adopting broadband Internet access at home. These responses are not surprising. Since dial-up users likely have already incurred the fixed costs of computers and equipment, the monthly subscription costs outweighed the fixed costs of computers and installation as the primary obstacle to home broadband service.

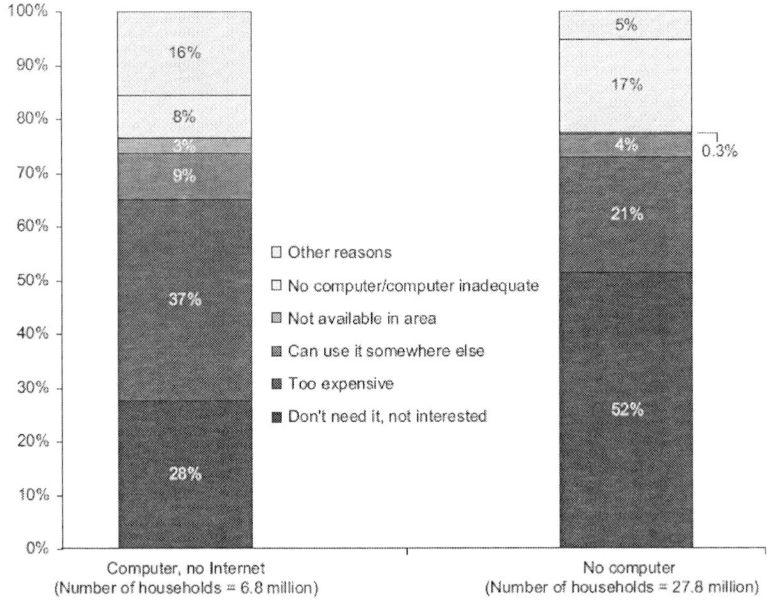

Source: U.S. Census Bureau, Current Population Survey School Enrollment and Internet Use Supplement, October 2010, and ESA calculations.

Figure 20. Main Reason for Not Having Home Internet Access Service, by Household Computer Ownership, 2010.

Among households that reported affordability as their main reason for not having any Internet access, both the fixed costs of purchasing a computer and equipment, as well as the monthly subscription costs were important (shown on the right panel of Figure 21). One-third of these households (33 percent) reported both monthly subscription and fixed costs, slightly more than one-fourth (27 percent) reported monthly cost alone, and almost one-third (30 percent) reported the fixed cost of a computer and equipment as their primary obstacle to having Internet access.

6. Internet Use Outside the Home, 2010

This report demonstrates that persistent differences exist in the adoption rate for broadband Internet access services across demographic groups, and that socio-economic and geographic factors, as well as demand or need, affordability, and availability all play significant roles in household adoption decisions. The variation in broadband Internet access across demographic groups points to an important area of investigation – do people who lack home broadband Internet access services use the Internet at locations outside the home and, if so, where do they access it? The expanded questions of the October 2010 CPS allow us to take a closer look at this topic.

Section 3 showed that more than two-thirds of American households (68 percent) utilized home broadband Internet access in 2010. According to Figure 22, the majority of these broadband households also used the Internet outside the home. Forty-three percent of American households had broadband Internet access at home and used the Internet outside the home, while 25 percent of households had broadband at home but did not use the Internet outside the home. Three percent of American households used dial-up at home – almost two percent of households had dial-up and did not use the Internet outside the home, and one percent of households both used dial-up and found additional locations outside the home from which they connected to the Internet.

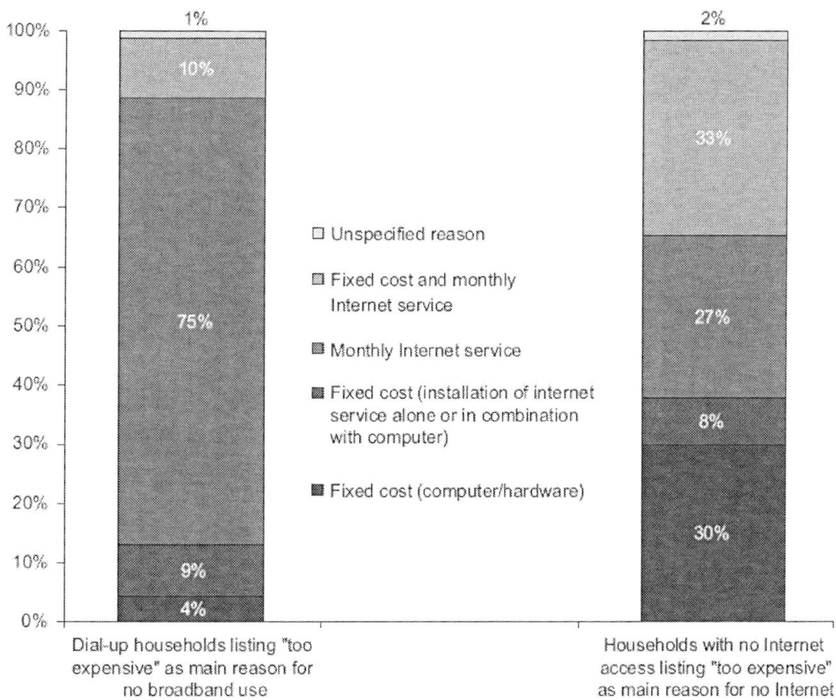

Source: U.S. Census Bureau, Current Population Survey School Enrollment and Internet Use Supplement, October 2010, and ESA calculations.

Figure 21. Detailed Reasons Why Internet Access Service Was Too Expensive.

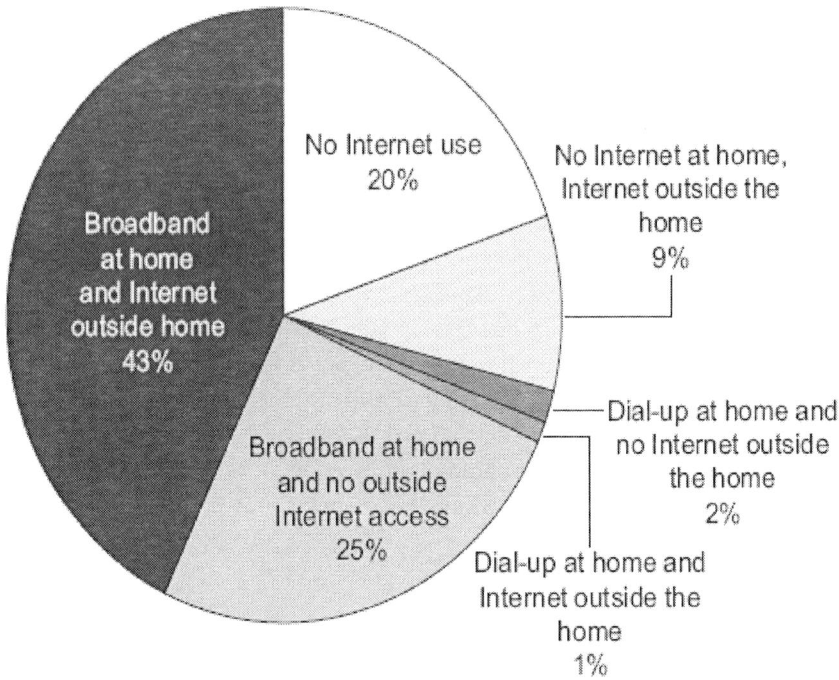

Source: U.S. Census Bureau, Current Population Survey School Enrollment and Internet Use Supplement, October 2010, and ESA calculations.

Figure 22. Household Distribution of Internet Access Points, 2010.

Figure 23 shows the tabulation results for locations outside of the home where people reported accessing the Internet. In general, those without broadband Internet access at home depended more heavily on public libraries or someone else's house for Internet access than those with broadband service. The left bar of Figure 23 shows that residents of households with no broadband primarily accessed the Internet at work, school, or the public library.

Those with broadband Internet access at home (showed by the right bar) primarily connected to the Internet at work or school, but relied less heavily on public libraries or someone else's house.

Only four percent of broadband households reported accessing the Internet at a public library, compared to 20 percent of those without broadband Internet access at home.

About two percent of broadband households also accessed the Internet at someone else's house, compared to 12 percent of non-broadband households.

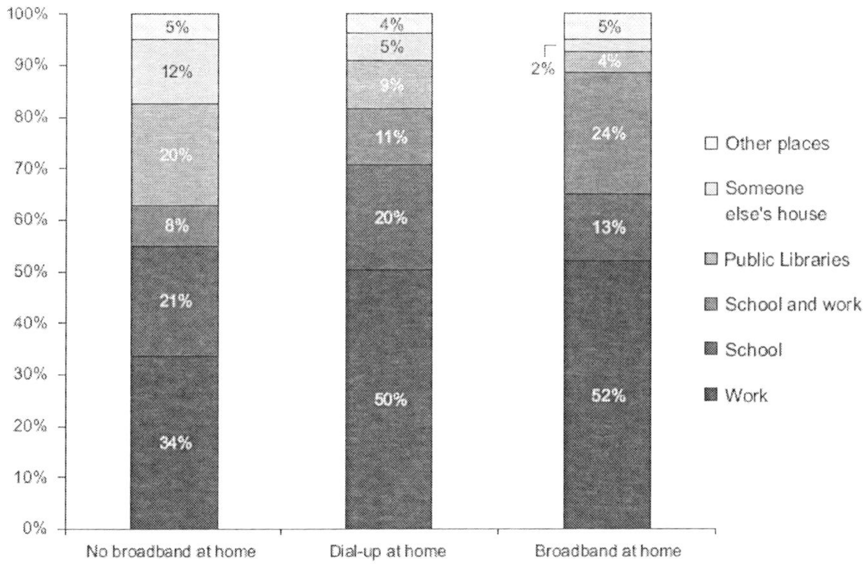

Source: U.S. Census Bureau, Current Population Survey School Enrollment and Internet Use Supplement, October 2010, and ESA calculations.

Note: Respondents could select up to seven different locations where they accessed the Internet outside of home. We developed mutually exclusive categories for Figure 23. "Work" includes households that accessed the Internet at work only, as well as households that accessed the Internet at work and some other location (other than school). "School" includes households that accessed the Internet at school only or in combination with some other location (other than work). "Public libraries" include households that accessed the Internet at public libraries only or in combination with locations other than work or school. "Someone else's house" includes households who accessed the Internet at someone else's house only or in combination with other places aside from work, school, or a public library. "Other places" include community centers, Internet cafés/coffee shops, or unspecified places.

Figure 23. Internet Access Outside the Home, by Home Broadband Adoption, 2010.

CONCLUSION

President Obama is dedicated to "connecting every part of America to the digital age" (NTIA, 2011). For policymakers, that commitment means ensuring that households have access to broadband services to give them the opportunity to participate in the information economy. As demonstrated in this report, while broadband adoption continues to grow, disparities persist. Households in certain demographic and geographic groups are less likely than their peers to use broadband services and, more importantly, these gaps remain to varying degrees even after socio-economic and geographic factors have been taken into account. The data and analyses provide policymakers with empirical evidence to help them address such disparities and find ways to enable all Americans to embrace the digital age.

APPENDIX A: DATA AND METHODOLOGY

This report uses data from the October 2010 Current Population Survey (CPS), a monthly survey of a representative sample of the U.S. noninstitutional population that provides data on labor force participation, income, and demographic characteristics of households. In addition, this report analyzes data from the most recent (October 2010) CPS School Enrollment and Internet Use Supplement, a special supplement to the CPS that periodically gathers information on Internet use.[21]

The October 2010 School Enrollment and Internet Use Supplement asked each surveyed household whether someone in that household used or owned a computer, used the Internet, and the locations from which they accessed the Internet (home, office, school, library, and other places). In addition, the survey asked the household which of the following technologies members utilized to connect to the Internet from home: dial-up service, DSL, cable modem, fiber optics, satellite, mobile broadband, or some other Internet connection technology. Using these data, it is possible to determine whether a household owned a personal computer (i.e., desktop, laptop, netbook, or notebook) or a handheld device, as well as the type of broadband technology (i.e., DSL, cable modem, fiber optics, satellite, or mobile broadband) members utilized to connect to the Internet. The survey also asked those households that did not access Internet services to state their main reason for not doing so.

Our sample consists of all households where the head of the household or "householder" is at least 16 years of age.[22] About 54,300 household records comprise our sample, representing 119.5 million American households. We analyze computer ownership and broadband Internet use at the household level and its association with household-level characteristics, such as income, size and composition, and geographic location.[23] For characteristics like education, race, ethnicity, age, disability status, and foreign-born status, we use information for the householder. In this report we use the words "adoption," "use," "utilization," and "access" interchangeably to indicate that a household reported having Internet access.

Data on computer use, as well as the types of broadband technology that online households utilized, have not been available since the early 2000s. The supply and demand for both mobile devices and residential Internet access services have changed enormously during this period. The data from the October 2010 CPS make it possible to identify the preferred or most common types of computers and access technologies used for residential Internet access, including the prevalence of mobile broadband technologies and handheld devices.

APPENDIX B: STATISTICAL TABLES

Table B1. Computer Ownership by Demographic Characterisitics, 2010

	Total	Desktop, laptop, netbook or notebook only (no handheld)	Handheld mobile device alone or in combination	No Computer
All households (standard error)	119,516,167 (328,675)	57.8% (0.2%)	18.3% (0.2%)	23.3% (0.2%)
	Household Income			
Less than $25,0000	34,509,443 (245,038)	45.1% (0.4%)	8.9% (0.2%)	45.6% (0.4%)
$25,000 to $50,000	32,961,706	62.1%	13.5%	24.0%
	(240,331)	(0.4%)	(0.3%)	(0.3%)
$50,000 to $75,000	21,228,025	66.5%	20.8%	12.2%
	(197,966)	(0.5%)	(0.4%)	(0.3%)
$75,000 to $100,000	12,472,081	66.6%	25.9%	6.9%
	(154,594)	(0.6%)	(0.6%)	(0.3%)
$100,000 or more	18,344,912	58.3%	36.5%	4.0%
	(185,178)	(0.5%)	(0.5%)	(0.2%)
	Householder Education			
No High School Diploma	14,150,347 (157,219)	38.0% (0.6%)	6.2% (0.3%)	55.6% (0.6%)
High School Diploma	34,946,796	55.0%	11.6%	33.1%
	(235,012)	(0.4%)	(0.2%)	(0.3%)
Some College	34,167,814	64.1%	19.0%	16.4%
	(232,836)	(0.4%)	(0.3%)	(0.3%)

Table B1. (Continued)

	Total	Desktop, laptop, netbook or notebook only (no handheld)	Handheld mobile device alone or in combination	No Computer
	Householder Education			
College Degree or More	36,251,210	62.5%	28.9%	7.7%
	(238,567)	(0.3%)	(0.3%)	(0.2%)
	Householder Race and Ethnicity			
White, Non-Hispanic	83,598,462	60.1%	19.3%	20.0%
	(315,569)	(0.2%)	(0.2%)	(0.2%)
Black, Non-Hispanic	14,853,701	50.2%	14.3%	35.1%
	(130,599)	(0.5%)	(0.4%)	(0.5%)
American Indian and	730,295	53.4%	12.2%	34.4%
Alaska Native, Non-Hispanic	(34,336)	(2.4%)	(1.6%)	(2.3%)
Asian, Non-Hispanic	4,667,055	61.2%	24.8%	13.6%
	(76,205)	(0.9%)	(0.8%)	(0.7%)
Hispanic	14,137,933	52.2%	14.0%	33.5%
	(159,699)	(0.7%)	(0.5%)	(0.7%)
	Householder Sex			
Male	60,041,712	58.5%	20.0%	20.8%
	(287,815)	(0.3%)	(0.2%)	(0.2%)
Female	59,474,455	57.2%	16.6%	25.8%
	(286,923)	(0.3%)	(0.2%)	(0.2%)
	Householder Age			
Average Age (years)	49.9	49.4	42.3	57.4
	(0.3)	(0.3)	(.05)	(0.6)
	Householder Disability Status			
Has a disability	16,381,088	45.2%	8.3%	46.1%
	(168,292)	(0.5%)	(0.3%)	(0.5%)
Does not have a disability	102,746,723	59.9%	19.8%	19.7%
	(326,579)	(0.2%)	(0.2%)	(0.2%)
	Household Geographic Location			
Metropolitan area	99,413,612	57.8%	19.7%	21.9%
	(325,338)	(0.2%)	(0.2%)	(0.2%)
Non-metropolitan area	19,236,331	58.0%	11.4%	30.3%
	(221,876)	(0.6%)	(0.4%)	(0.6%)

Source: U.S. Census Bureau, Current Population Survey School Enrollment and Internet Use Supplement, October 2010, and ESA calculations.

Note: Data for Whites, Blacks, Asians, and American Indians and Alaska Natives do not include people of Hispanic origin. Persons of Hispanic origin may be of any race.

Table B2. Computer Use, Internet and Broadband Adoption at the Household Level, by Demographic Characteristics and Geographic Locations, 2010

	Total	Computer use	Internet use	Broadband Adoption	
		All Households	All Households	All Households	Computer-using Households
All households	119,516,167	76.7%	71.1%	68.2%	88.9%
(standard error)	(328,675)	(0.2%)	(0.2%)	(0.2%)	(0.1%)
	Household Income				
Less than $25,0000	34,509,443	54.4%	45.9%	42.9%	78.8%
	(245,038)	(0.4%)	(0.4%)	(0.4%)	(0.4%)
$25,000 to $50,000	32,961,706	76.1%	69.3%	65.8%	86.6%
	(240,331)	(0.3%)	(0.4%)	(0.4%)	(0.3%)
$50,000 to $75,000	21,228,025	87.8%	83.7%	80.7%	91.9%
	(197,966)	(0.3%)	(0.4%)	(0.4%)	(0.3%)
$75,000 to $100,000	12,472,081	93.1%	90.1%	87.8%	94.4%
	(154,594)	(0.3%)	(0.4%)	(0.4%)	(0.3%)
$100,000 or more	18,344,912	96.0%	94.1%	92.6%	96.4%
	(185,178)	(0.2%)	(0.2%)	(0.3%)	(0.2%)
	Householder Education				
No High School Diploma	14,150,347	44.5%	35.7%	33.1%	74.6%
	(157,219)	(0.6%)	(0.5%)	(0.5%)	(0.7%)
High School Diploma	34,946,796	66.9%	60.3%	56.9%	85.0%
	(235,012)	(0.3%)	(0.4%)	(0.4%)	(0.3%)
Some College	34,167,814	83.6%	77.4%	74.3%	88.9%
	(232,836)	(0.3%)	(0.3%)	(0.3%)	(0.3%)
College Degree or More	36,251,210	92.3%	89.2%	87.2%	94.5%
	(238,567)	(0.2%)	(0.2%)	(0.2%)	(0.2%)
	Householder Race and Ethnicity				
White, Non-Hispanic	83,598,462	80.0%	74.9%	71.8%	89.7%
	(315,569)	(0.2%)	(0.2%)	(0.2%)	(0.2%)
Black, Non-Hispanic	14,853,701	64.9%	57.8%	55.5%	85.5%
	(130,599)	(0.5%)	(0.5%)	(0.5%)	(0.5%)
American Indian and Alaska Native, Non-Hispanic	730,295	65.6%	56.8%	52.3%	79.6%
	(34,336)	(2.3%)	(2.4%)	(2.4%)	(2.4%)
Asian, Non-Hispanic	4,667,055	86.4%	82.8%	80.9%	93.7%
	(76,205)	(0.7%)	(0.7%)	(0.7%)	(0.5%)
Hispanic	14,137,933	66.6%	59.1%	56.9%	85.5%
	(159,699)	(0.7%)	(0.7%)	(0.7%)	(0.6%)
	Householder Sex				
Male	60,041,712	79.2%	74.0%	71.2%	89.9%
	(287,815)	(0.2%)	(0.2%)	(0.3%)	(0.2%)
Female	59,474,455	74.2%	68.1%	65.3%	87.9%
	(286,923)	(0.2%)	(0.3%)	(0.3%)	(0.2%)

Table B2. (Continued)

	Householder Age				
16 to 44 years	47,979,027	84.4%	78.3%	76.7%	90.8%
	(266,130)	(0.2%)	(0.3%)	(0.3%)	(0.2%)
45 to 64 years	46,162,699	80.4%	75.3%	72.0%	89.5%
	(262,325)	(0.3%)	(0.3%)	(0.3%)	(0.2%)
65 years and older	25,374,441	55.4%	49.6%	45.5%	82.1%
	(205,050)	(0.4%)	(0.4%)	(0.4%)	(0.4%)
	Household Type				
Households with school-age children	29,996,963	86.3%	80.4%	78.1%	90.5%
	(220,445)	(0.3%)	(0.3%)	(0.3%)	(0.2%)
Households without school-age children	89,519,204	73.5%	67.9%	64.9%	88.3%
	(319,994)	(0.2%)	(0.2%)	(0.2%)	(0.2%)
	Householder Disability Status				
Has a disability	16,381,088	54.0%	46.4%	43.1%	79.8%
	(168,292)	(0.5%)	(0.5%)	(0.5%)	(0.6%)
Does not have a	102,746,723	80.3%	74.9%	72.2%	89.9%
	(326,579)	(0.2%)	(0.2%)	(0.2%)	(0.1%)
	Household				
Citizens (including	111,737,231	77.3%	71.8%	68.9%	89.1%
	(328,557)	(0.2%)	(0.2%)	(0.2%)	(0.1%)
Non-citizen	7,778,937	68.0%	60.7%	58.3%	85.7%
	(134,834)	(0.8%)	(0.9%)	(0.9%)	(0.7%)
	Urban-Rural				
Metropolitan Area	99,413,612	78.1%	72.8%	70.4%	90.1%
	(325,338)	(0.2%)	(0.2%)	(0.2%)	(0.2%)
Non-metropolitan Area	19,236,331	69.7%	62.0%	57.3%	82.1%
	(221,876)	(0.6%)	(0.6%)	(0.6%)	(0.6%)
	Metropolitan				
Less than 1,000,000	32,045,595	76.5%	70.8%	67.9%	88.9%
	(277,640)	(0.4%)	(0.4%)	(0.4%)	(0.3%)
1,000,000 to 2,499,999	20,735,483	77.5%	72.2%	69.8%	90.0%
	(187,425)	(0.4%)	(0.4%)	(0.4%)	(0.3%)
2,500,000 to	20,540,004	82.2%	77.3%	75.0%	91.2%
	(186,626)	(0.4%)	(0.4%)	(0.4%)	(0.3%)
5,000,000 or more	22,272,392	77.5%	73.0%	71.2%	91.8%
	(193,541)	(0.4%)	(0.4%)	(0.4%)	(0.4%)
Sample Size	54,269				
Population	119,516,167				

Source: U.S. Census Bureau, Current Population Survey School Enrollment and Internet Use Supplement, October 2010, and ESA calculations.

Note: Data for Whites, Blacks, Asians, and American Indians and Alaska Natives do not include people of Hispanic origin. Persons of Hispanic origin may be of any race.

Table B3. Computer Use, Internet and Broadband Adoption at the Individual Person Level, by Demographic Characteristics and Geographic Location, 2010

	Total	Computer use	Internet use	Broadband Adoption
All people age 3 and older	292,065,057	81.4%	65.0%	63.0%
(standard error)	(298,963)	(0.2%)	(0.2%)	(0.2%)
	Household Income			
Less than $25,0000	71,475,861 (472,342)	59.6% (0.4%)	40.6% (0.4%)	38.4% (0.4%)
$25,000 to $50,000	77,222,903 (482,761)	79.3% (0.3%)	60.0% (0.4%)	57.4% (0.4%)
$50,000 to $75,000	54,694,545 (432,720)	89.6% (0.3%)	73.5% (0.4%)	71.3% (0.4%)
$75,000 to $100,000	34,663,881	93.9%	80.1%	78.3%
	(362,174)	(0.3%)	(0.4%)	(0.5%)
$100,000 or more	54,007,868	96.9%	86.2%	85.1%
	(430,771)	(0.2%)	(0.3%)	(0.3%)
	Education			
No High School Diploma	41,927,111	64.2%	43.9%	42.1%
	(271,156)	(0.3%)	(0.4%)	(0.4%)
High School Diploma	71,043,770	73.4%	56.5%	53.7%
	(325,529)	(0.2%)	(0.3%)	(0.3%)
Some College	65,686,166	87.1%	77.0%	74.4%
	(318,033)	(0.2%)	(0.2%)	(0.2%)
College Degree or More	63,715,654	93.3%	87.1%	85.4%
	(315,025)	(0.1%)	(0.2%)	(0.2%)
	Race and Ethnicity			
White, Non-Hispanic	190,266,571	85.3%	71.4%	69.1%
	(590,558)	(0.2%)	(0.2%)	(0.2%)
Black, Non-Hispanic	35,096,831	70.2%	52.5%	50.7%
	(321,363)	(0.6%)	(0.7%)	(0.7%)
American Indian and Alaska Native, Non-Hispanic	1,750,907	68.7%	51.2%	47.1%
	(104,476)	(2.9%)	(3.1%)	(3.1%)
Asian, Non-Hispanic	13,413,301	90.3%	70.6%	69.3%
	(197,574)	(0.7%)	(1%)	(1%)

Table B3. (Continued)

	Total	Computer use	Internet use	Broadband Adoption
Hispanic	46,368,897	71.3%	47.5%	46.0%
	(189,554)	(0.7%)	(0.8%)	(0.8%)
	Sex			
Male	142,927,723	82.0%	65.0%	63.1%
	(602,891)	(0.2%)	(0.3%)	(0.3%)
Female	149,137,334	80.8%	65.1%	62.9%
	(604,465)	(0.2%)	(0.3%)	(0.3%)
	Age			
3 to 15 years	53,674,951	84.8%	53.0%	51.8%
	(456,464)	(0.3%)	(0.5%)	(0.5%)
16 to 44 years	118,971,154	85.7%	75.3%	73.5%
	(587,627)	(0.2%)	(0.3%)	(0.3%)
45 to 64 years	80,541,611	82.7%	69.2%	66.6%
	(529,334)	(0.3%)	(0.4%)	(0.4%)
65 years and older	38,877,341	61.0%	41.6%	38.5%
	(399,433)	(0.5%)	(0.5%)	(0.5%)
	Household Type			
Households with school-age children	122,429,615	86.4%	67.2%	65.5%
	(590,756)	(0.2%)	(0.3%)	(0.3%)
Households without school-age children	169,635,442	77.8%	63.5%	61.1%
	(602,859)	(0.2%)	(0.3%)	(0.3%)
	Disability Status			
Has a disability	26,615,113	60.6%	40.7%	38.3%
	(337,816)	(0.6%)	(0.7%)	(0.6%)
Does not have a disability	215,757,587	83.3%	71.3%	69.0%
	(559,451)	(0.2%)	(0.2%)	(0.2%)
	Citizenship Status			
Citizens (including foreign-born)	271,072,746	82.2%	66.3%	64.2%
	(410,258)	(0.2%)	(0.2%)	(0.2%)
Non-citizen	20,992,311	71.1%	48.6%	47.1%
	(345,420)	(0.8%)	(0.9%)	(0.9%)
	Urban-Rural Status			
Metropolitan Area	244,464,515	82.4%	66.4%	64.7%
	(499,361)	(0.2%)	(0.2%)	(0.2%)

	Total	Computer use	Internet use	Broadband Adoption
Non-metropolitan Area	45,509,222	75.8%	57.6%	53.9%
	(522,832)	(0.5%)	(0.6%)	(0.6%)
	Metropolitan Area (CBSA) Size			
Under 1,000,000	78,090,127	80.9%	65.1%	63.0%
	(641,721)	(0.4%)	(0.5%)	(0.5%)
1,000,000 to 2,499,999	49,326,079	82.2%	67.2%	65.5%
	(441,243)	(0.4%)	(0.5%)	(0.5%)
2,500,000 to 499,999,999	51,180,606	85.8%	70.2%	68.4%
	(447,875)	(0.3%)	(0.4%)	(0.4%)
5,000,000 and over	56,773,520	82.0%	64.7%	63.5%
	(466,636)	(0.3%)	(0.4%)	(0.4%)
Sample Size	129,494			
Population	292,065,057			

Source: U.S. Census Bureau, Current Population Survey School Enrollment and Internet Use Supplement, October 2010, and ESA calculations.

Note: Data for Whites, Blacks, Asians, and American Indians and Alaska Natives do not include people of Hispanic origin. Persons of Hispanic origin may be of any race.

Table B4. Marginal Effects: Regression Broadband Adoption on Demographic and Geographic Characteristics, 2010

Linear probability models	Having home broadband		Having home broadband(among computer owners)	
	(1)	(2)	(1)	(2)
Household Income: Less than $25,0000	Omitted	Omitted	Omitted	Omitted
Household Income: $25,000-$50,000	0.1368***	0.1366***	0.0582***	0.0581***
	(0.006)	(0.006)	(0.006)	(0.006)
Household Income: $50,000-$75,000	0.2156***	0.2153***	0.0927***	0.0926***
	(0.007)	(0.007)	(0.006)	(0.006)
Household Income: $75,000-$100,000	0.2403***	0.2395***	0.1029***	0.1023***
	(0.007)	(0.007)	(0.006)	(0.006)
Household Income: $100,000 or more	0.2460***	0.2441***	0.1083***	0.1072***
	(0.007)	(0.007)	(0.006)	(0.006)

Table B4. (Continued)

Linear probability models	Having home broadband		Having home broadband(among computer owners)	
	(1)	(2)	(1)	(2)
Education: No High School Diploma	Omitted	Omitted	Omitted	Omitted
Education: High School Diploma	0.1320***	0.1317***	0.0746***	0.0745***
	(0.008)	(0.008)	(0.010)	(0.010)
Education: Some College	0.2465***	0.2460***	0.1001***	0.0999***
	(0.008)	(0.008)	(0.010)	(0.010)
Education: College Degree or More	0.3048***	0.3035***	0.1262***	0.1254***
	(0.008)	(0.008)	(0.010)	(0.010)
Age	0.0093***	0.0094***	0.0031***	0.0031***
	(0.001)	(0.001)	(0.001)	(0.001)
Age squared	-0.0001***	-0.0001***	-0.0000***	-0.0000***
	(0)	(0)	(0.000)	(0.000)
White, Non-Hispanic	Omitted	Omitted	Omitted	Omitted
Black, Non-Hispanic	-0.1099***	-0.1126***	-0.0280***	-0.0299***
	(0.007)	(0.007)	(0.007)	(0.007)
Hispanic	-0.1096***	-0.1113***	-0.0281***	-0.0294***
	(0.008)	(0.008)	(0.007)	(0.007)
Asian, Non-Hispanic	-0.0052	-0.0078	0.0080	0.0063
	(0.01)	(0.01)	(0.007)	(0.007)
Other, Non-Hispanic	-0.0529***	-0.0534***	-0.0337**	-0.0337**
	(0.014)	(0.014)	(0.014)	(0.014)
Disability	-0.0634***	-0.0629***	-0.0328***	-0.0323***
	(0.007)	(0.007)	(0.007)	(0.007)
Disability-not identified	0.1065***	0.1074***	0.0636***	0.0650***
	(0.02)	(0.02)	(0.010)	(0.010)
Foreign-born non-citizen	-0.0473***	-0.0487***	-0.0208**	-0.0217**
	(0.01)	(0.01)	(0.009)	(0.009)
Total number of persons in household	0.0470***	0.0472***	0.0129***	0.0130***
	(0.002)	(0.002)	(0.002)	(0.002)
Has related school-age children	0.1217***	0.1224***	0.0174	0.0182
	(0.014)	(0.014)	(0.012)	(0.012)
Total number of persons in household for households	-0.0396***	-0.0398***	-0.0085***	-0.0087***
with related school-age children (interaction)	(0.004)	(0.004)	(0.003)	(0.003)
Rural (Nonmetropolitan)	Omitted	Omitted	Omitted	Omitted
Urban (Metropolitan)	0.0528***		0.0490***	
	(0.006)		(0.006)	
Metropolitan Area not identified	0.0617***		0.0211	
	(0.023)		(0.022)	
Metropolitan size: Less than		0.0492***		0.0464***

Linear probability models	Having home broadband		Having home broadband(among computer owners)	
	(1)	(2)	(1)	(2)
1,000,000				
		(0.007)		(0.006)
Metropolitan size: 1,000,000 to 2,499,999		0.0529***		0.0543***
		(0.008)		(0.007)
Metropolitan size: 2,500,000 to 499,999,999		0.0676***		0.0555***
		(0.008)		(0.008)
Metropolitan size: 5,000,000 or more		0.0661***		0.0593***
		(0.009)		(0.008)
Metropolitan area size not identified		0.0287***		0.0234**
		(-0.011)		(0.010)
Constant	0.0641**	0.0665**	0.5704***	0.5712***
	(0.026)	(0.026)	(0.028)	(0.028)
Sample Size	54,269	54,269	41,844	41,844
Estimated Number of Households	119,516,167	119,516,167	91,702,117	91,702,117
R-squared	0.276	0.277	0.073	0.073

Source: U.S. Census Bureau, Current Population Survey School Enrollment and Internet Use Supplement, October 2010, and ESA calculations.

Note: Standard errors in parentheses, *** $p<0.01$, ** $p<0.05$, * $p<0.1$. The marginal effects presented in the report are from the first and third columns. The columns labeled (2) include controls for urban (metropolitan) area size. State of residence is included in the regression. Data for Whites, Blacks, and Asians do not include people of Hispanic origin. Persons of Hispanic origin may be of any race.

Table B5. Internet and Broadband Adoption by State, 2010

		Broadband Adoption			Dial-up Noly		
			90 Percent Confidence Internval			90 Percent Confidence Internval	
State	Total Household	Percent	Lower Bound	Upper Bound	Percent	Lower Bound	Upper Bound
Alabama	1,924,324	55.5%	52.9%	58.2%	4.5%	3.4%	5.6%
Alaska	266,132	73.4%	70.8%	75.9%	5.3%	4.0%	6.6%
Arizona	2,552,907	74.1%	72.1%	76.2%	1.3%	0.8%	1.8%
Arkansas	1,205,789	52.4%	49.7%	55.1%	6.4%	5.1%	7.7%
California	12,932,685	73.1%	72.2%	74.0%	2.7%	2.4%	3.1%
Colorado	2,051,111	71.6%	69.2%	74.0%	3.1%	2.2%	4.1%
Connecticut	1,355,754	74.8%	72.3%	77.3%	1.7%	0.9%	2.4%
Delaware	347,772	68.4%	65.7%	71.0%	3.4%	2.3%	4.4%
District of Columbia	288,297	71.7%	69.2%	74.2%	1.7%	1.0%	2.4%

Table B5. (Continued)

State	Total Household	Broadband Adoption			Dial-up Noly		
		Percent	90 Percent Confidence Internval		Percent	90 Percent Confidence Internval	
			Lower Bound	Upper Bound		Lower Bound	Upper Bound
Florida	7,631,514	70.2%	69.0%	71.4%	1.8%	1.5%	2.2%
Georgia	3,846,149	68.6%	66.8%	70.4%	1.8%	1.3%	2.3%
Hawaii	461,347	69.2%	66.5%	71.8%	1.9%	1.1%	2.7%
Idaho	584,409	72.0%	69.5%	74.4%	3.6%	2.5%	4.6%
Illinois	5,057,694	68.7%	67.2%	70.3%	2.0%	1.6%	2.5%
Indiana	2,589,632	58.9%	56.6%	61.1%	2.4%	1.7%	3.2%
Iowa	1,229,080	67.4%	64.8%	70.1%	3.2%	2.2%	4.3%
Kansas	1,152,298	74.6%	72.2%	77.1%	1.7%	1.0%	2.5%
Kentucky	1,750,335	57.8%	55.0%	60.6%	3.5%	2.5%	4.6%
Louisiana	1,755,129	60.5%	57.7%	63.2%	2.3%	1.5%	3.2%
Maine	545,320	67.4%	64.4%	70.3%	6.0%	4.5%	7.5%
Maryland	2,214,385	74.1%	71.9%	76.4%	2.2%	1.5%	3.0%
Massachusetts	2,501,199	76.0%	73.9%	78.0%	1.6%	1.0%	2.2%
Michigan	3,956,678	66.3%	64.5%	68.1%	3.4%	2.7%	4.1%
Minnesota	2,135,427	70.6%	68.2%	72.9%	3.1%	2.2%	4.0%
Mississippi	1,135,683	51.7%	48.8%	54.5%	6.0%	4.6%	7.3%
Missouri	2,401,597	64.4%	62.0%	66.7%	3.5%	2.6%	4.4%
Montana	440,582	61.4%	58.8%	64.0%	4.0%	2.9%	5.0%
Nebraska	712,266	68.9%	66.3%	71.6%	2.3%	1.5%	3.2%
Nevada	1,012,500	74.2%	71.7%	76.7%	2.4%	1.5%	3.2%
New Hampshire	526,105	77.8%	75.3%	80.3%	3.2%	2.1%	4.2%
New Jersey	3,221,652	73.2%	71.4%	75.1%	1.5%	1.0%	2.0%
New Mexico	809,399	57.7%	54.9%	60.5%	4.9%	3.7%	6.1%
New York	7,739,363	69.0%	67.7%	70.3%	2.0%	1.7%	2.4%
North Carolina	3,674,129	65.1%	3.1%	-1.9%	4.2%	3.4%	5.0%
North Dakota	274,856	70.9%	68.3%	73.4%	2.2%	1.4%	3.1%
Ohio	4,681,232	63.9%	62.2%	65.6%	3.6%	2.9%	4.2%
Oklahoma	1,505,684	62.5%	59.7%	65.2%	3.7%	2.7%	4.8%
Oregon	1,554,311	74.7%	72.2%	77.2%	3.6%	2.5%	4.7%
Pennsylvania	5,129,874	67.4%	65.8%	68.9%	2.9%	2.3%	3.4%
Rhode Island	429,097	70.8%	68.1%	73.4%	1.3%	0.7%	2.0%
South Carolina	1,804,505	59.5%	56.8%	62.3%	4.3%	3.1%	5.4%
South Dakota	331,836	65.5%	63.0%	68.0%	3.5%	2.5%	4.5%
Tennessee	2,562,953	59.5%	57.2%	61.8%	3.8%	2.9%	4.7%
Texas	8,997,268	66.9%	65.7%	68.1%	2.7%	2.3%	3.1%
Utah	950,913	79.7%	77.5%	81.8%	2.6%	1.8%	3.5%
Vermont	263,979	69.2%	66.4%	72.0%	5.5%	4.1%	6.8%
Virginia	2,935,158	69.5%	67.5%	71.5%	3.5%	2.7%	4.3%

State	Total Household	Broadband Adoption			Dial-up Noly		
		Percent	90 Percent Confidence Internval		Percent	90 Percent Confidence Internval	
			Lower Bound	Upper Bound		Lower Bound	Upper Bound
Washington	2,781,539	76.7%	74.8%	78.6%	3.0%	2.2%	3.8%
West Virginia	737,127	59.1%	56.5%	61.7%	6.0%	4.7%	7.2%
Wisconsin	2,339,106	70.5%	68.3%	72.8%	3.2%	2.3%	4.0%
Wyoming	228,089	72.9%	70.4%	75.5%	1.5%	0.8%	2.2%

Source: U.S. Census Bureau, Current Population Survey School Enrollment and Internet Use Supplement, October 2010, and ESA calculations.

Table B6. Computer Use by State, 2010

States	Total Households	Computer, no Internet			No Computer		
			90 Percent Confidence		Percent	90 Percent Confidence	
			Lower Bound	Upper Bound		Lower Bound	Upper Bound
Alabama	1,924,324	8.5%	7.0%	10.0%	31.5%	29.0%	34.0%
Alaska	266,132	6.9%	5.4%	8.4%	14.4%	12.4%	16.5%
Arizona	2,552,907	4.8%	3.8%	5.8%	19.8%	17.9%	21.7%
Arkansas	1,205,789	8.8%	7.3%	10.3%	32.5%	29.9%	35.0%
California	12,932,685	5.3%	4.8%	5.8%	18.8%	18.0%	19.7%
Colorado	2,051,111	4.6%	3.5%	5.7%	20.6%	18.5%	22.8%
Connecticut	1,355,754	4.5%	3.3%	5.8%	19.0%	16.7%	21.3%
Delaware	347,772	4.5%	3.3%	5.7%	23.8%	21.4%	26.3%
District of Columbia	288,297	3.0%	2.0%	4.0%	23.6%	21.2%	26.0%
Florida	7,631,514	4.6%	4.0%	5.2%	23.4%	22.2%	24.5%
Georgia	3,846,149	6.0%	5.1%	6.9%	23.6%	21.9%	25.2%
Hawaii	461,347	3.5%	2.4%	4.6%	25.4%	22.9%	27.9%
Idaho	584,409	5.8%	4.5%	7.1%	18.7%	16.5%	20.8%
Illinois	5,057,694	4.2%	3.5%	4.9%	25.0%	23.6%	26.5%
Indiana	2,589,632	8.2%	6.9%	9.5%	30.5%	28.4%	32.6%
Iowa	1,229,080	6.1%	4.7%	7.4%	23.3%	20.9%	25.7%
Kansas	1,152,298	4.8%	3.6%	6.1%	18.8%	16.6%	21.0%
Kentucky	1,750,335	7.0%	5.6%	8.5%	31.6%	29.0%	34.2%
Louisiana	1,755,129	5.6%	4.3%	6.9%	31.6%	29.0%	34.2%
Maine	545,320	5.6%	4.2%	7.0%	21.1%	18.5%	23.6%
Maryland	2,214,385	3.7%	2.7%	4.7%	20.0%	17.9%	22.0%
Massachusetts	2,501,199	2.7%	1.9%	3.5%	19.7%	17.8%	21.6%
Michigan	3,956,678	6.6%	5.7%	7.6%	23.6%	22.0%	25.2%
Minnesota	2,135,427	5.5%	4.3%	6.6%	20.9%	18.8%	23.0%
Mississippi	1,135,683	9.8%	8.1%	11.5%	32.5%	29.9%	35.2%
Missouri	2,401,597	8.2%	6.9%	9.6%	24.0%	21.9%	26.1%
Montana	440,582	10.2%	8.6%	11.8%	24.5%	22.2%	26.8%

Table B6. (Continued)

States	Total Households	Computer, no Internet			No Computer		
			90 Percent Confidence			90 Percent Confidence	
			Lower Bound	Upper Bound	Percent	Lower Bound	Upper Bound
Nebraska	712,266	7.6%	6.1%	9.1%	21.2%	18.8%	23.5%
Nevada	1,012,500	4.3%	3.2%	5.5%	19.1%	16.9%	21.3%
New Hampshire	526,105	5.1%	3.8%	6.4%	13.9%	11.8%	15.9%
New Jersey	3,221,652	3.4%	2.6%	4.2%	21.9%	20.1%	23.6%
New Mexico	809,399	8.9%	7.3%	10.5%	28.5%	25.9%	31.0%
New York	7,739,363	4.0%	3.5%	4.6%	24.9%	23.7%	26.1%
North Carolina	3,674,129	4.5%	3.6%	5.3%	27.1%	25.4%	28.9%
North Dakota	274,856	4.2%	3.1%	5.3%	22.7%	20.3%	25.0%
Ohio	4,681,232	6.8%	5.9%	7.7%	25.7%	24.2%	27.3%
Oklahoma	1,505,684	6.3%	4.9%	7.6%	27.5%	25.0%	30.0%
Oregon	1,554,311	6.3%	4.9%	7.7%	15.4%	13.3%	17.4%
Pennsylvania	5,129,874	4.7%	4.0%	5.4%	25.1%	23.7%	26.6%
Rhode Island	429,097	5.3%	3.9%	6.6%	22.7%	20.2%	25.2%
South Carolina	1,804,505	6.4%	5.0%	7.7%	29.9%	27.3%	32.4%
South Dakota	331,836	7.2%	5.8%	8.5%	23.8%	21.6%	26.0%
Tennessee	2,562,953	7.0%	5.8%	8.2%	29.7%	27.6%	31.9%
Texas	8,997,268	7.2%	6.6%	7.9%	23.2%	22.2%	24.3%
Utah	950,913	4.4%	3.3%	5.5%	13.3%	11.4%	15.1%
Vermont	263,979	5.7%	4.3%	7.1%	19.6%	17.2%	22.0%
Virginia	2,935,158	6.8%	5.7%	7.9%	20.2%	18.5%	22.0%
Washington	2,781,539	6.8%	5.7%	8.0%	13.5%	11.9%	15.0%
West Virginia	737,127	8.7%	7.2%	10.2%	26.2%	23.8%	28.5%
Wisconsin	2,339,106	4.7%	3.7%	5.7%	21.6%	19.6%	23.6%
Wyoming	228,089	7.1%	5.6%	8.5%	18.6%	16.3%	20.8%

Source: U.S. Census Bureau, Current Population Survey School Enrollment and Internet Use Supplement, October 2010, and ESA calculations.

Table B7. Urban Area Broadband Adoption by State, 2010

States	Total Households	Broadband Adoption		
			90 Percent Confidence Interval	
		Percent	Lower Bound	Upper Bound
Alaska	182,990	76.9%	73.9%	79.9%
Arizona	2,238,063	75.6%	73.4%	77.8%
Arkansas	774,343	56.1%	52.8%	59.4%
California	12,670,211	73.1%	72.2%	74.0%
Colorado	1,801,148	71.2%	68.6%	73.8%
Connecticut	1,289,322	75.1%	72.5%	77.7%
Delaware	273,617	70.9%	67.9%	73.9%
District of Columbia	288,297	71.7%	69.2%	74.2%

		Broadband Adoption		
			90 Percent Confidence Interval	
States	Total Households	Percent	Lower Bound	Upper Bound
Florida	7,287,051	70.7%	69.4%	72.0%
Georgia	3,269,179	72.1%	70.2%	74.0%
Hawaii	325,395	71.1%	68.0%	74.2%
Idaho	365,668	78.2%	75.3%	81.1%
Illinois	4,407,181	70.3%	68.7%	71.9%
Indiana	1,862,332	60.5%	57.8%	63.2%
Iowa	688,449	73.0%	69.6%	76.4%
Kansas	707,374	79.2%	76.3%	82.1%
Kentucky	961,503	65.1%	61.5%	68.7%
Louisiana	1,505,052	62.5%	59.6%	65.4%
Maine	273,340	73.7%	69.8%	77.6%
Maryland	2,126,207	74.5%	72.2%	76.8%
Massachusetts	2,431,918	75.6%	73.5%	77.7%
Michigan	3,307,369	67.6%	65.7%	69.5%
Minnesota	1,556,618	72.4%	69.7%	75.1%
Mississippi	469,103	66.8%	62.6%	71.0%
Missouri	1,867,125	68.6%	66.0%	71.2%
Montana	149,610	69.1%	64.9%	73.3%
Nebraska	434,971	75.0%	71.8%	78.2%
Nevada	867,609	75.4%	72.8%	78.0%
New Hampshire	321,621	81.2%	78.2%	84.2%
New Jersey	3,221,652	73.2%	71.3%	75.1%
New Mexico	592,076	61.4%	58.2%	64.6%
New York	7,085,604	69.5%	68.2%	70.8%
North Carolina	2,440,415	69.0%	66.8%	71.2%
North Dakota	129,583	78.3%	75.0%	81.6%
Ohio	3,589,034	66.0%	64.1%	67.9%
Oklahoma	1,034,283	68.1%	64.9%	71.3%
Oregon	1,181,748	77.3%	74.6%	80.0%
Pennsylvania	4,215,756	67.7%	66.0%	69.4%
Rhode Island	429,097	70.7%	68.0%	73.4%
South Carolina	1,213,534	63.9%	60.6%	67.2%
South Dakota	161,230	71.9%	68.5%	75.3%
Tennessee	1,877,190	64.5%	61.8%	67.1%
Texas	7,968,051	67.8%	66.5%	69.1%
Utah	729,619	79.5%	77.0%	81.9%
Vermont	82,160	81.1%	76.9%	85.3%
Virginia	2,526,293	72.9%	70.8%	75.0%
Washington	2,566,102	77.4%	75.4%	79.4%
West Virginia	442,531	63.9%	60.6%	67.2%
Wisconsin	1,765,333	72.4%	69.9%	74.9%
Wyoming	65,142	77.5%	73.0%	82.0%

Source: U.S. Census Bureau, Current Population Survey School Enrollment and Internet Use Supplement, October 2010, and ESA calculations.

Table B8. Rural Area Broadband Adoption by State, 2010

State	Total Households	Broadband Adoption		
			90 Percent Confidence Interval	
		Percent	Lower Bound	Upper Bound
Alabama	529,809	47.0%	40.8%	53.2%
Alaska	83,143	65.5%	59.4%	71.6%
Arizona	314,844	63.6%	55.7%	71.5%
Arkansas	431,446	45.6%	40.1%	51.1%
California	262,475	72.8%	64.7%	80.9%
Connecticut	66,432	69.0%	54.1%	83.9%
Delaware	74,155	59.0%	51.5%	66.5%
Florida	344,463	59.9%	52.3%	67.5%
Georgia	576,970	48.9%	42.9%	54.9%
Hawaii	135,952	64.4%	58.1%	70.7%
Idaho	218,741	61.6%	56.3%	66.9%
Illinois	650,512	58.4%	52.8%	64.0%
Indiana	727,300	54.5%	49.2%	59.8%
Iowa	540,631	60.3%	55.2%	65.4%
Kansas	444,925	67.3%	62.0%	72.6%
Kentucky	788,832	49.0%	43.8%	54.2%
Maine	271,980	61.0%	55.7%	66.3%
Maryland	88,178	63.6%	48.4%	78.8%
Michigan	649,310	59.8%	54.2%	65.4%
Minnesota	578,809	65.5%	59.8%	71.2%
Mississippi	666,580	41.1%	36.6%	45.6%
Missouri	534,472	49.4%	43.0%	55.8%
Montana	290,972	57.4%	53.4%	61.4%
Nebraska	277,295	59.3%	53.7%	64.9%
New Hampshire	204,484	72.5%	67.3%	77.7%
New Mexico	217,324	47.6%	41.0%	54.2%
New York	653,759	63.8%	58.2%	69.4%
North Carolina	1,233,714	57.6%	53.5%	61.7%
North Dakota	145,273	64.3%	59.8%	68.8%
Ohio	1,092,197	57.1%	52.7%	61.5%
Oklahoma	471,401	50.0%	43.9%	56.1%
Oregon	372,564	66.5%	59.7%	73.2%
Pennsylvania	914,118	65.7%	61.1%	70.3%
South Carolina	590,971	50.5%	44.5%	56.4%
South Dakota	170,606	59.5%	55.1%	63.8%
Tennessee	685,763	45.8%	40.3%	51.3%
Texas	1,029,217	59.9%	55.4%	64.4%
Vermont	181,819	63.8%	59.5%	68.1%
Virginia	408,865	48.5%	41.3%	55.7%
Washington	215,436	68.4%	59.1%	77.7%

State	Total Households	Broadband Adoption		
			90 Percent Confidence Interval	
		Percent	Lower Bound	Upper Bound
West Virginia	294,595	52.0%	46.9%	57.1%
Wisconsin	573,773	64.6%	58.8%	70.4%
Wyoming	162,947	71.1%	67.3%	74.9%

Source: U.S. Census Bureau, Current Population Survey School Enrollment and Internet Use Supplement, October 2010, and ESA calculations.

Note: The following states were not included in this figure due to data limitations for rural areas: Colorado, Louisiana, Nevada, Rhode Island, Massachusetts, New Jersey and Utah. The District of Columbia had no rural areas.

REFERENCES

Bureau of Labor Statistics and U.S. Bureau of the Census. "CPS Record Layout for Basic Labor Force Items, Standard Public Use Files." Current Population Survey, January 2010, Accessed June 12, 2011. *http://smpbff2.dsd.census.gov/pub/cps/basic/201001-/jan10dd.txt*

Economics and Statistics Administration and National Telecommunications and Information Administration. "Exploring the Digital Nation: Home Broadband Internet Adoption in the United States." November 2010, Accessed May 26, 2011. *http://www.esa.doc.gov/Reports/ exploring-digital-nation-home-broadband-Internet-adoption-united-states*

Federal Communications Commission. "Broadband Adoption and Use in America." OBI Working Paper Series No. 1, February 2010, Accessed September 7, 2011. *http://hraunfoss.fcc.gov/edocs_public/attachmatch/ DOC-296442A1.pdf*

National Telecommunications and Information Administration. "Digital Nation: Expanding Internet Usage." February 2011, Accessed May 26, 2011. *http://www.ntia.doc.gov/report/2011/ digital-nation-expanding-internet-usage-ntia-research-preview*

National Telecommunications and Information Administration. Digital Nation: 21st Century America's Progress Toward Universal Broadband Internet Access, February 2010, Accessed June 15, 2011. *http://www.ntia.doc.gov /files/ntia/publications/ntia_internet_use_report_february_2011.pdf*

Organisation for Economic Co-operation and Development (OECD). OECD Broadband Statistics. "Table 1d (2): OECD Terrestrial mobile wireless broadband subscriptions per 100 inhabitants, by technology, June 2010."

Accessed May 26, 2011. *http://www.oecd.org/dataoecd/21/35/ 39574709. xls*

Office of Management and Budget (OMB). OMB Bulletin No. 04-03, "Update of Statistical Area Definitions and Additional Guidance on Their Uses." Accessed June 23, 2010. *http://www.whitehouse.gov/omb/bulletins_fy 04_b04-03/*

Smith, Aaron. "Mobile Access 2010." Pew Internet and American Life Project, Pew Research Center, July 7, 2010, Accessed September 7, 2011. *http://www.pewinternet.org/Reports/2010/Mobile-Access-2010.aspx*

U.S. Bureau of the Census. "School Enrollment and Internet Use Supplement File Technical Documentation, CPS-09." Current Population Survey, October 2009, Accessed September 7, 2011. *http://www.census.gov/ apsd/techdoc/cps/cpsoct09.pdf*

U.S. Bureau of the Census (a). "Metropolitan and Micropolitan Statistical Areas." Accessed September 7, 2011. *http://www.census.gov/population /www/metroareas/ aboutmetro.html*

White House Office of the Press Secretary. Remarks by the President on Fiscal Policy in Washington, DC April 13, 2011, Accessed September 21, 2011. *http://www.whitehouse.gov/the-press-office/2011/04/13/remarks-president-fiscal-policy*

End Notes

[1] In this report, we examine Internet access service from the demand side based on the Census Bureau's survey of households. We use the terms "adoption," "use," "utilization," "access," and "connection" interchangeably to indicate that a household reported having Internet access service. The term "Internet access service" includes both the provision of dial-up Internet access service and broadband Internet access service. Similarly, the CPS survey inquires about households' ownership or use of a home computer to examine whether they have access to the devices necessary to access the Internet. The report describes such access as "ownership" or "use" and employs the terms interchangeably.

[2] Please see Appendix A for a detailed description of the data and methodology employed in this report.

[3] For the 2010 household-level estimate of computer usage based on the total sample, the margin of error at the 90 percent confidence level is plus or minus 0.35 percentage points based on a standard error (SE) of 0.21 percentage points. See Appendix B, Table B2 for estimates of computer use and standard errors for population subgroups. The last time the CPS included data on computer use was in 2003.

[4] All comparisons referenced in this paper have been tested for statistical significance, and are significant at the 90 percent confidence level.

[5] The householder is the person (or one of the people) who owns or rents the housing unit where the subject household resides, or, if there is no such person, any adult member, excluding roomers, boarders, or paid employees.

[6] A household with at least one of the following high-speed, high capacity, two-way Internet services is considered to have broadband: DSL, cable modem, fiber optics, satellite, mobile broadband, or some other non-dial-up Internet connection. The CPS did not ask about the speed of the particular broadband service a household uses because of the difficulty of gathering the information. The household-level estimate on broadband Internet access based on the total sample has a margin of error of plus or minus 0.38 percentage points, based on a standard of error of 0.23 percentage points. See Appendix B, Table B2 for estimates of Internet access and standard errors for population subgroups.

[7] The shares of households with DSL, cable modem, fiber optics, and satellite technologies shown in Figure 3 include households that reported having only that specific type of technology. The vast majority of households with these technologies had only one type of Internet technology. In contrast, the six percent of households with mobile broadband includes households that reported having mobile broadband, either alone or in addition to other types of technologies. The category "Other broadband services" includes unspecified technology types, or any other combination of listed technology types excluding mobile broadband. For example, the total share of households with DSL was 24.6 percent, including 23.4 percent with DSL alone, 0.8 percent with DSL and mobile broadband, and 0.4 percent with DSL and other broadband service. Similarly, the total share of households with cable modem was 33.9 percent, with 32.0 percent with cable modem alone, 1.4 percent with cable modem and mobile broadband, and 0.5 percent with cable modem and other broadband service.

[8] For purposes of the CPS, mobile broadband does not refer to Wi-Fi networks, which ultimately rely on a different transmission mode (e.g., one might have a cable modem at home that is connected to a Wi-Fi base station for wireless Internet access at home, but from the provider's perspective it is a cable connection). It also does not include satellite connections.

[9] Note that Figure 3 refers only to mobile broadband adoption at home. This differs from our finding that 19 percent of households used a handheld device (see Figure 2), which alone does not indicate mobile broadband use inside the home. The higher mobile broadband adoption rates reported in some other studies measure activities that are not strictly limited to the use of a mobile broadband device at home. For example, an OECD study found mobile broadband subscriptions to be 44.4 per 100 people in the United States (OECD, 2010). This number includes both home and "business" subscriptions. Also, a Pew study found that 40 percent of American adults use their cell phones to access the Internet, email, or instant messaging, regardless of whether the cell phones are used inside or outside the home (Smith, 2010). Similarly, a working paper from the Federal Communications Commission (FCC) found that 30 percent of American adults used their cell phones to send emails, access web pages, or download applications (FCC, 2010).

[10] The geographic variable for identifying a household's location as urban or rural is not available in the CPS public use files. This report uses the terms "urban" and "rural" to refer to metropolitan and nonmetropolitan areas, respectively. The definition of a metropolitan area (effective since 2000) is based on "core based statistical area" (CBSA), which includes both metropolitan and micropolitan statistical areas. According to the 2000 standards, each CBSA must have at least one urban area with at least 10,000 inhabitants. Each metropolitan statistical area must contain at least one urbanized area with a population of 50,000 or more.

Each micropolitan statistical area must contain at least one urban cluster with population of between 10,000 and 50,000. As of June 6, 2003, there are 362 metropolitan statistical areas and 560 micropolitan statistical areas in the United States. For more information, see U.S. Census Bureau (2010a) and Office of Management and Budget (2010).

[11] These percentages are based on calculations set forth in Appendix B, Table B2.

[12] Percentages in this paragraph are based on calculations of the actual numbers (see Appendix B, Table B2) and may not precisely equal the sum of the percentages shown in the accompanying figure due to rounding.

[13] As described in the Data and Methodology section (Appendix A), the data on race and ethnicity (as well as education, age, disability status, and foreign-born status) are for the householder. As a result, "White households" (for example) refer to households headed by a White person. The same definition applies to Black, Hispanic, Asian, and American Indian or Alaska Native households. Data for Whites, Blacks, Asians, and American Indians and Alaska Natives do not include people of Hispanic origin. Persons of Hispanic origin may be of any race.

[14] Percentages in this paragraph are based on calculations of the actual numbers (see Appendix B, Table B2) and may not precisely equal the sum of the percentages shown in the accompanying figure due to rounding.

[15] Figure 9 does not include data on American Indian and Alaska Native householders due to data limitations for this group by Internet connection technology.

[16] In the CPS, a civilian adult is considered to have a disability if he or she reported having at least one of the following conditions: (1) hearing impairment; (2) blindness or impaired vision despite wearing glasses; (3) physical, mental, or emotional condition that impairs the ability to concentrate, remember, or make decisions; (4) difficulty walking or climbing stairs; (5) difficulty dressing or bathing; or (6) physical, mental, or emotional condition that impairs the ability to do errands alone, such as visiting a doctor's office or shopping (Bureau of Labor Statistics and U.S. Census Bureau, 2010).

[17] Note that states are ordered by estimated average household broadband usage rate for ease of understanding and not as a specific ranking. Rates for broadband, Internet, and computer use should be understood in the context of their associated confidence intervals, set forth in Appendix B, Tables B5 and B6.

[18] Id. See Appendix B, Tables B7, B8.

[19] Household characteristics include income, education, age, race, disability status, citizenship status, presence of children, and population density. See Table B4 in Appendix B for more detail.

[20] The results presented in Appendix B, Table B4 show the urban-rural gap by urban area size.

[21] This report analyzes data from the most recent survey conducted in October 2010, the ninth such Internet survey conducted since the early 1990s. For a more detailed description of the survey, see Bureau of Labor Statistics and U.S. Census Bureau, 2010.

[22] The householder is the person (or one of the people) who owns or rents the housing unit, or, if there is no such person, any adult member, excluding roomers, boarders, or paid employees.

[23] Detailed data on computer use and broadband adoption both at the household and individual level are presented in Appendix B, Tables B1, B2, and B3.

In: Explorations of Internet Usage and Access ... ISBN: 978-1- 61942-923-9
Editors: B. Foster and M. Gonzales

Chapter 2

BROADBAND INTERNET ACCESS AND THE DIGITAL DIVIDE: FEDERAL ASSISTANCE PROGRAMS*

Lennard G. Kruger and Angele A. Gilroy

SUMMARY

The "digital divide" is a term that has been used to characterize a gap between "information haves and have-nots," or in other words, between those Americans who use or have access to telecommunications and information technologies and those who do not. One important subset of the digital divide debate concerns high-speed Internet access and advanced telecommunications services, also known as broadband. Broadband is provided by a series of technologies (e.g., cable, telephone wire, fiber, satellite, wireless) that give users the ability to send and receive data at volumes and speeds far greater than traditional "dial-up" Internet access over telephone lines.

Broadband technologies are currently being deployed primarily by the private sector throughout the United States. While the numbers of new broadband subscribers continue to grow, studies and data suggest that the rate of broadband deployment in urban/suburban and high income areas are outpacing deployment in rural and low-income areas. Some

---

* This is an edited, reformatted and augmented version of a Congressional Research Service publication, CRS Report for Congress RL30719 Prepared for Members and Committees of Congress, from www.crs.gov, dated October 24, 2011.

policymakers, believing that disparities in broadband access across American society could have adverse economic and social consequences on those left behind, assert that the federal government should play a more active role to avoid a "digital divide" in broadband access.

With the conclusion of the grant and loan awards announced by broadband programs temporarily established by the American Recovery and Reinvestment Act of 2009 (P.L. 111-5), the Rural Broadband Access Loan and Loan Guarantee Program and the Community Connect Broadband Grants, both at the Rural Utilities Service of the U.S. Department of Agriculture, are currently the only ongoing federal funding programs exclusively dedicated to deploying broadband infrastructure. However, there exist other federal programs that provide financial assistance for various aspects of telecommunications development. The major vehicle for funding telecommunications development, particularly in rural and low-income areas, is the Universal Service Fund (USF). While the USF's High Cost Program, as currently designed, does not explicitly fund broadband infrastructure, subsidies are used, in many cases, to upgrade existing telephone networks so that they are capable of delivering high-speed services. Additionally, subsidies provided by USF's Schools and Libraries Program and Rural Health Care Program are used for a variety of telecommunications services, including broadband access.

To the extent that the 112th Congress may consider various options for further encouraging broadband deployment and adoption, a key issue is how to strike a balance between providing federal assistance for unserved and underserved areas where the private sector may not be providing acceptable levels of broadband service, while at the same time minimizing any deleterious effects that government intervention in the marketplace may have on competition and private sector investment.

INTRODUCTION

The "digital divide" is a term used to describe a perceived gap between "information haves and have-nots," or in other words, between those Americans who use or have access to telecommunications and information technologies and those who do not.[1] Whether or not individuals or communities fall into the "information haves" category depends on a number of factors, ranging from the presence of computers in the home, to training and education, to the availability of affordable Internet access.

Broadband technologies are currently being deployed primarily by the private sector throughout the United States. While the numbers of new broadband subscribers continue to grow, studies and data suggest that the rate of broadband deployment in urban/suburban and high income areas are outpacing deployment in rural and low-income areas.

STATUS OF BROADBAND DEPLOYMENT IN THE UNITED STATES

Prior to the late 1990s, American homes accessed the Internet at maximum speeds of 56 kilobits per second by dialing up an Internet Service Provider (such as AOL) over the same copper telephone line used for traditional voice service. A relatively small number of businesses and institutions used broadband or high speed connections through the installation of special "dedicated lines" typically provided by their local telephone company. Starting in the late 1990s, cable television companies began offering cable modem broadband service to homes and businesses. This was accompanied by telephone companies beginning to offer DSL service (broadband over existing copper telephone wireline). Growth has been steep, rising from 2.8 million high speed lines reported as of December 1999, to 169 million lines as of December 31, 2010.[2] Of the 168 million high speed lines reported by the FCC, 140 million serve residential users.[3] Table 1 depicts the relative deployment of different types of broadband technologies. A distinction is often made between "current generation" and "next generation" broadband (commonly referred to as next generation networks or NGN). "Current generation" typically refers to currently deployed cable, DSL, and many wireless systems, while "next generation" refers to dramatically faster download and upload speeds offered by fiber technologies and also potentially by future generations of cable, DSL, and wireless technologies.4 In general, the greater the download and upload speeds offered by a broadband connection, the more sophisticated (and potentially valuable) the application that is enabled.

Based on the latest FCC broadband connection data, *Table 2* shows the percentages of households with broadband connections by state, both for download connections over 200 kbps and for connections of at least 3 Mbps (which approximates the FCC's National Broadband Availability target). According to the FCC, high speed connections over 200 kbps are reported in 65% of households nationwide, while connections of at least 3 Mbps (download) and 768 kbps (upload) are reported in 35% of households nationwide.

Table 1. Percentage of Broadband Technologies by Types of Connection (as of June 30, 2010)

	Connections over 200 kbps in at least one direction	Residential connections over 200 kbps in at least one direction	Connections at least 3 Mbps downstream and 768 kbps upstream	Residential connections at least 3 Mbps downstream and 768 kbps upstream
cable modem	26.8%	30.9%	58.0%	64.2%
DSL	18.6%	19.6%	13.1%	13.4%
Mobile wireless	50.0%	45.0%	19.9%	12.7%
Fiber	3.0%	3.4%	8.5%	9.4%
All other[a]	1.6%	1.0%	0.6%	0.3%

Source: FCC, Internet Access Services: Status as of December 31, 2010, pp. 24-27.

[a] Includes satellite, fixed wireless, power line, and all copper-wire based technologies other than DSL.

Similarly, the U.S. Department of Commerce report, *Digital Nation: Expanding Internet Usage* (based on October 2010 U.S. Census Bureau survey data) found that 68% of U.S. households have adopted broadband.[5] According to the FCC's National Broadband Plan, 290 million Americans—95% of the U.S. population—currently live in housing units with access to terrestrial, fixed broadband infrastructure capable of supporting actual download speeds of at least 4 Mbps. This leaves a "gap" of 14 million people in the United States living in 7 million housing units that do not have access to terrestrial broadband infrastructure capable of this speed.[6]

Meanwhile, according to the National Broadband Map, mandated by the ARRA and released by the NTIA on February 17, 2011, 98.3% of the U.S. population has minimum advertised broadband speeds available (at least 768 kbps download/200 kbps upload), while 95.5% have available advertised speeds of at least 3 Mbps (download) and 768 kbps (upload).[7] According to NTIA, 5% to 10% of Americans lack access to broadband at the FCC actual speed benchmark of 4 Mbps (download) and 1 Mbps (upload).[8]

Thus, while the broadband *adoption* or *penetration* rate stands at about 65% of U.S. households, broadband *availability* is much higher, at 95% or more of households, which means that approximately 30% of households have access to some type of terrestrial (non-satellite) broadband service, but do not choose to subscribe. An FCC consumer survey, conducted in October and November 2009, found that 35% or 80 million American adults do not use broadband at home, falling into three categories: those who do not use the Internet at all (22%); those who use the Internet but do not have Internet access at home (6%); and those who use dial-up to access the Internet (6%). The survey identified three primary reasons why non-adopting Americans do not have broadband: cost, lack of digital literacy, and the perceived insufficient relevance of broadband.[9] Similarly, according to the Department of Commerce report, the two most common reasons cited for not having broadband at home are that it is perceived as not needed or too expensive. Lack of a home computer can also be a major factor.[10] The Department of Commerce report, the FCC's National Broadband Plan, and a survey conducted by the Pew Internet and American Life Project[11] also found disparities in broadband adoption among demographic groups. Populations continuing to lag behind in broadband adoption include people with low incomes, seniors, minorities, the less-educated, non-family households, and the non-employed.

Meanwhile, the FCC's *Seventh Broadband Deployment Report*, released on May 20, 2011, estimated that 26.2 million Americans living in more than 9.2 million households are unserved by broadband.[12]

Broadband in Rural Areas[13]

While the number of new broadband subscribers continues to grow, the rate of broadband deployment in urban areas appears to be outpacing deployment in rural areas. While there are many examples of rural communities with state of the art telecommunications facilities,[14] recent surveys and studies have indicated that, in general, rural areas tend to lag behind urban and suburban areas in broadband deployment.

- The Department of Commerce report, *Digital Nation*, found that while the digital divide between urban and rural areas has lessened since 2007, it still persists with 70% of urban households accessing broadband service in 2009, compared to 60% of rural households.[15]
- Data from the Pew Internet & American Life Project indicate that while broadband adoption is growing in rural areas, broadband users

make up larger percentages of non-rural users than rural users. Pew found that the percentage of all U.S. adults with broadband at home is 70% for non-rural areas and 50% for rural areas.[16]

Table 2. Percentage of Households With Broadband Connections by State (as of December 31, 2010)

	Connections over 200 kbps	Connections at least 3 mbps downstream and 768 kbps upstream
Alabama	53%	21%
Alaska	70%	*
Arizona	67%	40%
Arkansas	50%	18%
California	72%	41%
Colorado	71%	50%
Connecticut	75%	48%
Delaware	73%	61%
District of Columbia	67%	51%
Florida	71%	39%
Georgia	59%	30%
Hawaii	*	*
Idaho	58%	15%
Illinois	65%	33%
Indiana	60%	30%
Iowa	63%	16%
Kansas	65%	23%
Kentucky	56%	30%
Louisiana	56%	28%
Maine	69%	19%
Maryland	72%	60%
Massachusetts	75%	64%
Michigan	61%	36%
Minnesota	66%	39%
Mississippi	44%	11%
Missouri	59%	20%

	Connections over 200 kbps	Connections at least 3 mbps downstream and 768 kbps upstream
Montana	61%	30%
Nebraska	66%	32%
Nevada	63%	33%
New Hampshire	76%	52%
New Jersey	78%	69%
New Mexico	57%	29%
New York	72%	45%
North Carolina	64%	12%
North Dakota	67%	34%
Ohio	65%	17%
Oklahoma	56%	23%
Oregon	65%	46%
Pennsylvania	68%	46%
Rhode Island	72%	*
South Carolina	59%	17%
South Dakota	63%	38%
Tennessee	52%	29%
Texas	61%	23%
Utah	68%	44%
Vermont	70%	50%
Virginia	67%	54%
Washington	68%	51%
West Virginia	53%	27%
Wisconsin	65%	23%
Wyoming	62%	37%
National subscribership ratio	65%	35%

Source: FCC, Internet Access Services: Status as of December 31, 2010, pp. 35-36.

Notes: Asterisk (*) indicates data withheld by the FCC to maintain firm confidentiality. Subscribership ratio is the number of reported residential high speed lines (broadband connections) divided by the number of households in each state.

- According to data from the National Broadband Map, 98.3% of the population in urban areas have available broadband speeds of at least 3 Mbps (download)/768 kbps (upload), as opposed to 84.8% of the population in rural areas.[17]
- The FCC's Seventh Broadband Deployment Report found "on average, unserved Americans live in areas that are more rural and less densely populated."[18]

The comparatively lower population density of rural areas is likely the major reason why broadband is less deployed than in more highly populated suburban and urban areas. Particularly for wireline broadband technologies—such as cable modem and DSL—the greater the geographical distances among customers, the larger the cost to serve those customers. Thus, there is often less incentive for companies to invest in broadband in rural areas than, for example, in an urban area where there is more demand (more customers with perhaps higher incomes) and less cost to wire the market area.[19]

Some policymakers believe that disparities in broadband access across American society could have adverse consequences on those left behind, and that advanced telecommunications applications critical for businesses and consumers to engage in e-commerce are increasingly dependent on high speed broadband connections to the Internet. Thus, some say, communities and individuals without access to broadband could be at risk to the extent that connectivity becomes a critical factor in determining future economic development and prosperity. A February 2006 study done by the Massachusetts Institute of Technology for the Economic Development Administration of the Department of Commerce marked the first attempt to quantitatively measure the impact of broadband on economic growth. The study found that "between 1998 and 2002, communities in which mass-market broadband was available by December 1999 experienced more rapid growth in employment, the number of businesses overall, and businesses in IT-intensive sectors, relative to comparable communities without broadband at that time."[20]

A June 2007 report from the Brookings Institution found that for every one percentage point increase in broadband penetration in a state, employment is projected to increase by 0.2 to 0.3% per year. For the entire U.S. private non-farm economy, the study projected an increase of about 300,000 jobs.[21]

Subsequently, a July 2009 study commissioned by the Internet Innovation Alliance found net consumer benefits of home broadband on the order of $32 billion per year, up from an estimated $20 billion in consumer benefits from home broadband in 2005.[22]

Some also argue that broadband is an important contributor to U.S. future economic strength with respect to the rest of the world. Data from the Organization for Economic Cooperation and Development (OECD) found the U.S. ranking 15th among OECD nations in broadband access per 100 inhabitants as of December 2010.[23] By contrast, in 2001 an OECD study found the U.S. ranking fourth in broadband subscribership per 100 inhabitants (after Korea, Sweden, and Canada).[24] While many argue that declining U.S. performance in international broadband rankings is a cause for concern,[25] others maintain that the OECD data undercount U.S. broadband deployment,[26] and that cross-country broadband deployment comparisons are not necessarily meaningful and are inherently problematic.[27] Finally, an issue related to international broadband rankings is the extent to which broadband speeds and prices differ between the U.S. and the rest of the world.[28]

Is Broadband Deployment Data Adequate?

Obtaining an accurate snapshot of the status of broadband deployment is problematic. Anecdotes abound of rural and low-income areas which do not have adequate Internet access, as well as those which are receiving access to high-speed, state-of-the-art connections. Rapidly evolving technologies, the constant flux of the telecommunications industry, the uncertainty of consumer wants and needs, and the sheer diversity and size of the nation's economy and geography make the status of broadband deployment very difficult to characterize. The FCC periodically collects broadband deployment data from the private sector via "FCC Form 477"—a standardized information gathering survey. Statistics derived from the Form 477 survey are published every six months. Additionally, data from Form 477 are used as the basis of the FCC's (to date) six broadband deployment reports.

The FCC is working to refine the data used in future Reports in order to provide an increasingly accurate portrayal. In its March 17, 2004 Notice of Inquiry for the *Fourth Report*, the FCC sought comments on specific proposals to improve the FCC Form 477 data gathering program.[29] On November 9, 2004, the FCC voted to expand its data collection program by requiring reports from all facilities based carriers regardless of size in order to better track rural and underserved markets, by requiring broadband providers to provide more information on the speed and nature of their service, and by establishing broadband-over-power line as a separate category in order to track its development and deployment. The FCC Form 477 data gathering program was extended for five years beyond its March 2005 expiration date.[30]

On April 16, 2007, the FCC announced a Notice of Proposed Rulemaking which sought comment on a number of broadband data collection issues, including how to develop a more accurate picture of broadband deployment; gathering information on price, other factors determining consumer uptake of broadband, and international comparisons; how to improve data on wireless broadband; how to collect information on subscribership to voice over Internet Protocol service (VoIP); and whether to modify collection of speed tier information.[31]

On March 19, 2008, the FCC adopted an order that substantially expands its broadband data collection capability. Specifically, the order expands the number of broadband reporting speed tiers to capture more information about upload and download speeds offered in the marketplace, requires broadband providers to report numbers of broadband subscribers by census tract, and improves the accuracy of information collected on mobile wireless broadband deployment. Additionally, in a Further Notice of Proposed Rulemaking, the FCC sought comment on broadband service pricing and availability.[32] The July 2009 data release (providing data as of June 30, 2008) was the final data set gathered under the old FCC Form 477. The February 2010 data report (December 31, 2008 data) reflected the new Form 477 data collection requirements. Meanwhile, during the 110th Congress, state initiatives to collect broadband deployment data in order to promote broadband in underserved areas were viewed as a possible model for governmental efforts to encourage broadband. The Broadband Data Improvement Act was enacted by the 110th Congress and became P.L. 110-385 on October 10, 2008. The law requires the FCC to collect demographic information on unserved areas, data comparing broadband service with 75 communities in at least 25 nations abroad, and data on consumer use of broadband. The act also directs the Census Bureau to collect broadband data, the Government Accountability Office to study broadband data metrics and standards, and the Department of Commerce to provide grants supporting state broadband initiatives.

P.L. 111-5, the American Recovery and Reinvestment Act, provided NTIA with an appropriation of $350 million to implement the Broadband Data Improvement Act and to develop and maintain a national broadband inventory map. The National Broadband Map was released on February 17, 2011 (www.broadbandmap.gov) and will be updated every six months.[33]

Finally, the FCC's National Broadband Plan addressed the broadband data issue, recommending that the FCC and the U.S. Bureau of Labor Statistics (BLS) should collect more detailed and accurate data on actual availability, penetration, prices, churn, and bundles offered by broadband service providers to consumers and businesses, and should publish analyses of these data.

BROADBAND AND THE FEDERAL ROLE

The Telecommunications Act of 1996 (P.L. 104-104) addressed the issue of whether the federal government should intervene to prevent a "digital divide" in broadband access. Section 706 requires the FCC to determine whether "advanced telecommunications capability [i.e., broadband or high-speed access] is being deployed to all Americans in a reasonable and timely fashion."

Since 1999, the FCC has adopted and released seven reports pursuant to Section 706. The first five reports formally concluded that the deployment of advanced telecommunications capability to all Americans is reasonable and timely. Unlike the first five 706 reports, the sixth and seventh reports concluded that broadband is not being deployed to all Americans in a reasonable and timely fashion. According to the *Seventh Broadband Deployment Report*:

> Our analysis of the best data available – the data collected by the National Telecommunications and Information Administration (NTIA) for the National Broadband Map – shows that as many as 26 million Americans live in areas unserved by broadband capable of originating and receiving high-quality voice, data, graphics, and video telecommunications. Many of these Americans live in areas where there is no business case to offer broadband, and where existing public efforts to extend broadband are unlikely to reach; they have no immediate prospect of being served, despite the growing costs of digital exclusion. For these and other reasons, we must conclude that broadband is not being deployed in a reasonable and timely fashion to all Americans.[34]

FCC Commissioner Robert McDowell issued a dissenting statement, maintaining that there is insufficient justification for the 706 report conclusion that broadband is not being deployed in a reasonable and timely fashion.

Specifically, the dissent argued that the report did not sufficiently account for the dramatic growth in the availability and deployment of mobile broadband.[35]

Section 706 of the Telecommunications Act of 1996 states that if the FCC formally determines that broadband is not being deployed in a reasonable and timely fashion, the FCC is directed "take immediate action to accelerate deployment of such capability by removing barriers to infrastructure investment and by promoting competition in the telecommunications market." According to the *Seventh Broadband Deployment Report*, the FCC has already begun this action by initiating various proceedings to implement the National Broadband Plan.[36]

The National Broadband Plan

As mandated by the ARRA, on March 16, 2010, the FCC publically released its report, *Connecting America: The National Broadband Plan.*[37] The National Broadband Plan (NBP) seeks to "create a high-performance America" which the FCC defines as "a more productive, creative, efficient America in which affordable broadband is available everywhere and everyone has the means and skills to use valuable broadband applications."[38] In order to achieve this mission, the NBP recommends that the country set six goals for 2020:

1. Goal No. 1: At least 100 million U.S. homes should have affordable access to actual download speeds of at least 100 megabits per second and actual upload speeds of at least 50 megabits per second.
2. Goal No. 2: The United States should lead the world in mobile innovation, with the fastest and most extensive wireless networks of any nation.
3. Goal No. 3: Every American should have affordable access to robust broadband service, and the means and skills to subscribe if they so choose.
4. Goal No. 4: Every American community should have affordable access to at least 1 gigabit per second broadband service to anchor institutions such as schools, hospitals and government buildings.
5. Goal No. 5: To ensure the safety of the American people, every first responder should have access to a nationwide, wireless, interoperable broadband public safety network.

6. Goal No. 6: To ensure that America leads in the clean energy economy, every American should be able to use broadband to track and manage their real-time energy consumption.

The National Broadband Plan is categorized into three parts:

- *Part I (Innovation and Investment)* which "discusses recommendations to maximize innovation, investment and consumer welfare, primarily through competition. It then recommends more efficient allocation and management of assets government controls or influences."[39] The recommendations address a number of issues, including spectrum policy, improved broadband data collection, broadband performance standards and disclosure, special access rates, interconnection, privacy and cybersecurity, child online safety, poles and rightsof-way, research and experimentation (R&E) tax credits, R&D funding.
- *Part II (Inclusion)* which "makes recommendations to promote inclusion—to ensure that all Americans have access to the opportunities broadband can provide."[40] Issues include reforming the Universal Service Fund, intercarrier compensation, federal assistance for broadband in Tribal lands, expandingexisting broadband grant and loan programs at the Rural Utilities Service, enable greater broadband connectivity in anchor institutions, and improved broadband adoption and utilization especially among disadvantaged and vulnerable populations.
- *Part III (National Purposes)* which "makes recommendations to maximize the use of broadband to address national priorities. This includes reforming laws, policies and incentives to maximize the benefits of broadband in areas where government plays a significant role."[41] National purposes include health care, education, energy and the environment, government performance, civic engagement, and public safety. Issues include telehealth and health IT, online learning and modernizing educational broadband infrastructure, digital literacy and job training, smart grid and smart buildings, federal support for broadband in small businesses, telework within the federal government, cybersecurity and protection of critical broadband infrastructure, copyright of public digital media, interoperable public safety communications, next generation 911 networks and emergency alert systems.

The release of the National Broadband Plan is seen by many as a precursor towards the development of a national broadband policy—whether comprehensive or piecemeal—that will likely be shaped and developed by Congress, the FCC, and the Administration. Upon release of the NBP, President Obama issued the following statement:

> My Administration will build upon our efforts over the past year to make America's nationwide broadband infrastructure the world's most powerful platform for economic growth and prosperity, including improving access to mobile broadband, maximizing technology innovation, and supporting a nationwide, interoperable public safety wireless broadband network.[42]

Meanwhile, Congress will play a major role in implementing the National Broadband Plan, both by considering legislation to implement NBP recommendations, and by overseeing broadband activities conducted by the FCC and executive branch agencies.

Current Federal Broadband Programs

With the conclusion of grant and loan awards announced by the broadband programs temporarily established by the American Recovery and Reinvestment Act of 2009 (P.L. 111-5),[43] the Rural Broadband Access Loan and Loan Guarantee Program and the Community Connect Broadband Grants, both at the Rural Utilities Service of the U.S. Department of Agriculture, are currently the only ongoing federal funding programs *exclusively* dedicated to deploying broadband infrastructure. However, there exist other federal programs that provide financial assistance for various aspects of telecommunications development.

The major vehicle for funding telecommunications development, particularly in rural and low-income areas, is the Universal Service Fund (USF). While the USF's High Cost Program does not *explicitly* fund broadband infrastructure, subsidies are used, in many cases, to upgrade existing telephone networks so that they are capable of delivering high-speed services. Additionally, subsidies provided by USF's Schools and Libraries Program and Rural Health Care Program are used for a variety of telecommunications services, including broadband access.

Table 3 (at the end of this report) shows selected federal domestic assistance programs throughout the federal government that currently can be associated with broadband and telecommunications development. The table categorizes the programs in three ways: programs exclusively devoted to the deployment of broadband infrastructure; programs which focus on or include deployment of telecommunications infrastructure generally (which typically can and does include broadband); and applications-specific programs which fund some aspect of broadband access or adoption as a means towards supporting a particular application, such as distance learning or telemedicine.

Rural Utilities Service Programs

The Rural Electrification Administration (REA), subsequently renamed the Rural Utilities Service (RUS), was established by the Roosevelt Administration in 1935. Initially, it was established to provide credit assistance for the development of rural electric systems. In 1949, the mission of REA was expanded to include rural telephone providers. Congress further amended the Rural Electrification Act in 1971 to establish within REA a Rural Telephone Account and the Rural Telephone Bank (RTB). Rural Telephone Loans and Loan Guarantees provide long-term direct and guaranteed loans for telephone lines, facilities, or systems to furnish and improve telecommunications service in rural areas. The RTB—liquidated in FY2006—was a public-private partnership intended to provide additional sources of capital that would supplement loans made directly by RUS. Another program, the Distance Learning and Telemedicine Program, specifically addresses health care and education needs of rural America.

RUS implements two programs specifically targeted at providing assistance for broadband deployment in rural areas: the Rural Broadband Access Loan and Loan Guarantee Program and Community Connect Broadband Grants. The 110th Congress reauthorized and reformed the Rural Broadband Access Loan and Loan Guarantee program as part of the 2008 farm bill (P.L. 110- 234).

The Universal Service Concept and the FCC[44]

Since its creation in 1934 the Federal Communications Commission (FCC) has been tasked with "mak[ing] available, so far as possible, to all the people of the United States ... a rapid, efficient, Nation-wide, and world-wide wire and radio communications service with adequate facilities at reasonable charges."[45] This mandate led to the development of what has come to be known as the universal service concept.

The universal service concept, as originally designed, called for the establishment of policies to ensure that telecommunications services are available to all Americans, including those in rural, insular and high cost areas, by ensuring that rates remain affordable.

Over the years this concept fostered the development of various FCC policies and programs to meet this goal. The FCC offers universal service support through a number of direct mechanisms that target both providers of and subscribers to telecommunications services.[46]

The development of the federal universal service high cost fund is an example of provider-targeted support. Under the high cost fund, eligible telecommunications carriers, usually those serving rural, insular and high cost areas, are able to obtain funds to help offset the higher than average costs of providing telephone service.[47] This mechanism has been particularly important to rural America where the lack of subscriber density leads to significant costs. FCC universal service policies have also been expanded to target individual users. Such federal programs include two income-based programs, Link Up and Lifeline, established in the mid-1980s to assist economically needy individuals. The Link Up program assists low-income subscribers with paying the costs associated with the initiation of telephone service and the Lifeline program assists low-income subscribers with paying the recurring monthly service charges. Effective January 1, 1998, schools, libraries, and rural health care providers also qualified for universal service support.

Universal Service and the Telecommunications Act of 1996

Passage of the Telecommunications Act of 1996 (P.L. 104-104) codified the long-standing commitment by U.S. policymakers to ensure universal service in the provision of telecommunications services.

The Schools and Libraries and Rural Health Care Programs

Congress, through the 1996 act, not only codified, but also expanded the concept of universal service to include, among other principles, that elementary and secondary schools and classrooms, libraries, and rural health care providers have access to telecommunications services for specific purposes at discounted rates. (See §§254(b)(6) and 254(h)of the 1996 Telecommunications Act, 47 USC. 254.)

1. The Schools and Libraries Program. Under universal service provisions contained in the 1996 act, elementary and secondary schools and classrooms and libraries are designated as beneficiaries of universal service discounts. Universal service principles detailed in Section 254(b)(6) state that "Elementary and secondary schools and classrooms ... and libraries should have access to advanced telecommunications services." The act further requires in Section 254(h)(1)(B) that services within the definition of universal service be provided toelementary and secondary schools and libraries for education purposes at discounts, that is at "rates less than the amounts charged for similar services to other parties." The FCC established the Schools and Libraries Division within the Universal Service Administrative Company (USAC) to administer the schools and libraries or "E (education)- rate" program to comply with these provisions. Under this program, eligible schools and libraries receive discounts ranging from 20 to 90 percent for telecommunications services depending on the poverty level of the school's (or school district's) population and its location in a high cost telecommunications area. Three categories of services are eligible for discounts: internal connections (e.g., wiring, routers and servers); Internet access; and telecommunications and dedicated services, with the third category receiving funding priority. According to data released by program administrators, approximately $29 billion in funding has been committed over the first thirteen years of the program with funding released to all states, the District of Columbia and all territories. Funding commitments for funding Year 2011 (July 1, 2011-June 30, 2012), the 14th and current year of the program, totaled almost $ 1.1 billion as of October 18, 2011.[48]
2. The Rural Health Care Program. Section 254(h) of the 1996 act requires that public and non-profit rural health care providers have access to telecommunications services necessary for the provision of health care services at rates comparable to those paid for similar services in urban areas. Subsection 254(h)(1) further specifies that "to the extent technically feasible and economically reasonable" health care providers should have access to advanced telecommunications and information services. The FCC established the Rural Health Care

Division (RHCD) within the USAC to administer the universal support program to comply with these provisions. Under FCC established rules only public or non-profit health care providers are eligible to receive funding. Eligible health care providers, with the exception of those requesting only access to the Internet, must also be located in a rural area. The funding ceiling, or cap, for this support was established at $400 million annually. The funding level for Year One of the program (January 1998-June 30, 1999) was set at $100 million. Due to less than anticipated demand, the FCC established a $12 million funding level for the second year (July 1, 1999 to June 30, 2000) of the program but has since returned to a $400 million yearly cap. As of March 31, 2011, covering the first 13 years of the program, a total of $449.9 million has been committed to 5,015 rural health care providers. The primary use of the funding is to provide reduced rates for telecommunications and information services necessary for the provision of health care.[49] In addition, the FCC established, in 2007, the "Rural Health Care Pilot Program" to help public and non-profit health care providers build state and region-wide broadband networks dedicated to the provision of health care services. As of March 31, 2011, $145.2 million of the $417.8 million authorized for the Pilot program has been committed to 40 of the 62 projects designated by the FCC to receive support under the program.

Universal Service and Broadband

One of the policy debates surrounding universal service is whether access to advanced telecommunications services (i.e., broadband) should be incorporated into universal service objectives. The term universal service, when applied to telecommunications, refers to the ability to make available a basket of telecommunications services to the public, across the nation, at a reasonable price. As directed in the 1996 Telecommunications Act [Section 254(c)] a federal-state Joint Board was tasked with defining the services which should be included in the basket of services to be eligible for federal universal service support; in effect using and defining the term "universal service" for the first time. The Joint Board's recommendation, which was subsequently adopted by the FCC in May 1997, included the following in its universal service package: voice grade access to and some usage of the public switched network; single line service; dual tone signaling; access to directory assistance; emergency service such as 911; operator services; and access and interexchange (long distance) service.

Some policy makers expressed concern that the FCC-adopted definition is too limited and does not take into consideration the importance and growing acceptance of advanced services such as broadband and Internet access. They point to a number of provisions contained in the Universal Service section of the 1996 Act to support their claim. Universal service principles contained in Section 254(b)(2) state that "Access to advanced telecommunications services should be provided to all regions of the Nation." The subsequent principle (b)(3) calls for consumers in all regions of the nation including "low-income" and those in "rural, insular, and high cost areas" to have access to telecommunications and information services including "advanced services" at a comparable level and a comparable rate charged for similar services in urban areas. Such provisions, they state, dictate that the FCC expand its universal service definition.

The 1996 act does take into consideration the changing nature of the telecommunications sector and allows for the universal service definition to be modified if future conditions warrant. Section 254(c)of the act states that "universal service is an evolving level of telecommunications services" and the FCC is tasked with "periodically" reevaluating this definition "taking into account advances in telecommunications and information technologies and services."

Furthermore, the Joint Board is given specific authority to recommend "from time to time" to the FCC modification in the definition of the services to be included for federal universal service support. The Joint Board, on November 19, 2007, concluded such an inquiry and recommended that the FCC change the mix of services eligible for universal service support. The Joint Board recommended, among other things, that "the universal availability of broadband Internet services" be included in the nation's communications goals and hence be supported by federal universal service funds.[50] The FCC in its national broadband plan, *Connecting America: the National Broadband Plan,* recommended that access to and adoption of broadband be a national goal.

Furthermore the national broadband plan proposes that the Universal Service Fund be restructured to become a vehicle to help reach this goal and the FCC has initiated a series of proceedings to achieve this.[51] Others caution that a more modest approach is appropriate given the "universal mandate" associated with this definition and the uncertainty and costs associated with mandating nationwide deployment of such advanced services as a universal service policy goal.

LEGISLATION IN THE 110TH CONGRESS

In the 110th Congress, legislation was enacted to provide financial assistance for broadband deployment.

Of particular note is the reauthorization of the Rural Utilities Service (RUS) broadband loan program, which was enacted as part of the 2008 farm bill (P.L. 110-234).

In addition to reauthorizing and reforming the RUS broadband loan program, P.L. 110-234 contains provisions establishing a National Center for Rural Telecommunications Assessment and requiring the FCC and RUS to formulate a comprehensive rural broadband strategy.

The Broadband Data Improvement Act (P.L. 110-385) was enacted by the 110th Congress and required the FCC to collect demographic information on unserved areas, data comparing broadband service with 75 communities in at least 25 nations abroad, and data on consumer use of broadband. The act also directed the Census Bureau to collect broadband data, the Government Accountability Office to study broadband data metrics and standards, and the Department of Commerce to provide grants supporting state broadband initiatives.

Meanwhile, the America COMPETES Act (H.R. 2272) was enacted (P.L. 110-69) and contained a provision authorizing the National Science Foundation (NSF) to provide grants for basic research in advanced information and communications technologies. Areas of research included affordable broadband access, including wireless technologies. P.L. 110-69 also directs NSF to develop a plan that describes the current status of broadband access for scientific research purposes.

LEGISLATION IN THE 111TH CONGRESS

In the 111th Congress, legislation was introduced that sought to provide financial assistance for broadband deployment. Of particular note, provisions in the American Recovery and Reinvestment Act of 2009 (P.L. 111-5) provided grants and loans to support broadband access and adoption in unserved and underserved areas.

P.L. 111-5: The American Recovery and Reinvestment Act of 2009

On February 17, 2009, President Obama signed P.L. 111-5, the American Recovery and Reinvestment Act (ARRA). Broadband provisions of the ARRA provided a total of *$7.2 billion*, for broadband grants, loans, and loan/grant combinations.

The total consisted of $4.7 billion to NTIA/DOC for a newly established Broadband Technology Opportunities Program (grants) and $2.5 billion to the RUS/USDA Broadband Initiatives Program (grants, loans, and grant/loan combinations).[52]

Regarding the $2.5 billion to RUS/USDA broadband programs, the ARRA specified that at least 75% of the area to be served by a project receiving funds shall be in a rural area without sufficient access to high speed broadband service to facilitate economic development, as determined by the Secretary of Agriculture.

Priority was given to projects that provide service to the most rural residents that do not have access to broadband services. Priority was also given to borrowers and former borrowers of rural telephone loans.

Of the $4.7 billion appropriated to NTIA:

- $4.35 billion was directed to a competitive broadband grant program, of which not less than $200 million shall be available for competitive grants for expanding public computer center capacity (including at community colleges and public libraries); not less than $250 million to encourage sustainable adoption of broadband service; and $10 million transferred to the Department of Commerce Office of Inspector General for audits and oversight; and
- $350 million was directed for funding the Broadband Data Improvement Act (P.L. 110-385) and for the purpose of developing and maintaining a broadband inventory map, which shall be made accessible to the public no later than two years after enactment. Funds deemed necessary and appropriate by the Secretary of Commerce may be transferred to the FCC for the purposes of developing a national broadband plan, which shall be completed one year after enactment.

Final BTOP and BIP program awards were announced by September 30, 2010.

Other Broadband Legislation in the 111th Congress

P.L. 111-8 (H.R. 1105). Omnibus Appropriations Act, 2009. Appropriates to RUS/USDA $15.619 million to support a loan level of $400.487 million for the Rural Broadband Access Loan and Loan Guarantee Program, and $13.406 million for the Community Connect Grant Program. To the FCC, designates not less than $3 million to establish and administer a State Broadband Data and Development matching grants program for State-level broadband demand aggregation activities and creation of geographic inventory maps of broadband service to identify gaps in service and provide a baseline assessment of statewide broadband deployment. Passed House February 25, 2009. Passed Senate March 10, 2009. Signed by President, March 12, 2009.

P.L. 111-32 (H.R. 2346). Supplemental Appropriations Act, 2009. Provides not less than $3 million to the FCC to develop a national broadband plan pursuant to the American Recovery and Reinvestment Act of 2009. Introduced May 12, 2009; referred to Committee on Appropriations. Passed House May 14, 2009; passed Senate May 21, 2009. Signed by President, June 24, 2009.

P.L. 111-80 (H.R. 2997). Agriculture, Rural Development, Food and Drug Administration, and Related Agencies Appropriations Act, 2010. For Rural Utilities Service, U.S. Department of Agriculture, provides $28.96 million to support a loan level of $400 million for the broadband loan program, and $17.97 million for broadband community connect grants. Introduced June 23, 2009; referred to Committee on Appropriations. Reported by Committee on Appropriations June 23, 2009. Passed House July 9, 2009. Passed Senate August 4, 2009. Conference Report (H.Rept. 111-279) printed September 30, 2009. Signed by President October 21, 2009.

H.R. 691 (Meeks). Broadband Access Equality Act of 2009. Amends the Internal Revenue Code of 1986 to provide credit against income tax for businesses furnishing broadband services to underserved and rural areas. Introduced January 26, 2009; referred to Committee on Ways and Means.

H.R. 760 (Eshoo). Advanced Broadband Infrastructure Bond Initiative of 2009. Amends the Internal Revenue Code of 1986 to provide an income tax credit to holders of bonds financing new advanced broadband infrastructure. Introduced January 28, 2009; referred to Committee on Ways and Means and in addition to Committee on Energy and Commerce.

H.R. 2428 (Eshoo). Broadband Conduit Deployment Act of 2009. Directs the Secretary of Transportation to require that broadband conduit be installed as part of certain highway construction projects. Introduced May 14, 2009; referred to Committee on Transportation and Infrastructure.

H.R. 2521 (DeLauro). National Infrastructure Development Bank Act of 2009. Establishes a National Infrastructure Development Bank to finance infrastructure projects, including broadband and telecommunications projects. Introduced May 20, 2009; referred to Committee on Energy and Commerce and in addition to the Committees on Transportation and Infrastructure, and on Financial Services.

H.R. 3101 (Markey). Twenty-first Century Communications and Video Accessibility Act of 2009. Ensures that individuals with disabilities have access to emerging Internet Protocol-based communication and video program technologies in the 21st Century. Introduced June 26, 2009; referred to Committee on Energy and Commerce.

H.R. 3413 (Capito). Rural Information Technology Investment Act. Authorizes the National Telecommunications and Information Administration of the Department of Commerce to make grants for the establishment of information technology centers in rural areas. Introduced July 30, 2009; referred to Committee on Energy and Commerce.

H.R. 3646 (Matsui). Broadband Affordability Act of 2009. Amends the Communications Act of 1934 to establish a Lifeline Assistance Program for universal broadband adoption. Introduced September 24, 2009; referred to Committee on Energy and Commerce.

H.R. 4545 (Murphy). Rural Broadband Initiative Act of 2010. Establishes an Office of Rural Broadband Initiatives in the Department of Agriculture which would administer the RUS broadband loan and grant programs, and would develop a comprehensive rural broadband strategy. Establishes a National Rural Broadband Innovation Fund, authorized at $20 million for each of fiscal years 2008 through 2012, that would fund experimental and pilot rural broadband projects. Introduced January 27, 2010; referred to Committee on Agriculture and in addition to the Committee on Energy and Commerce.

H.R. 4619 (Markey). E-Rate 2.0 Act of 2010. Amends the Communications Act of 1934 to create a pilot program to bridge the digital divide by providing vouchers for broadband service to eligible students, to increase access to advanced telecommunications and information services for community colleges and head start programs, and to establish a pilot program for discounted electronic books. Introduced February 9, 2010; referred to Committee on Energy and Commerce.

H.R. 5828 (Boucher). Universal Service Reform Act of 2010. Reforms the universal service provisions of the Communications Act of 1934 and other purposes. Introduced July 22, 2010; referred to the Committee on Energy and Commerce.

S. 1266 (Klobuchar). Broadband Conduit Deployment Act of 2009. Directs the Secretary of Transportation to require that broadband conduit be installed as part of certain highway construction projects. Introduced June 15, 2009; referred to Committee on Environment and Public Works.

S. 1447 (Hutchison). Connecting America Act of 2009. Provides broadband Internet investment tax credits and credits to holders of broadband bonds. Also establishes an Office of National Broadband Strategy in the National Telecommunications and Information Administration and provides broadband adoption incentives in telehealth and distance learning programs. Introduced July 14, 2009; referred to Committee on Finance.

S. 2879 (Rockefeller). Broadband Opportunity and Affordability Act. Directs the FCC to conduct a pilot program expanding the Lifeline Program to include broadband service. Also directs the FCC to prepare a report exploring whether the Link Up program should be expanded to include computer ownership in order to reduce the cost of initiating broadband service. Introduced December 11, 2009; referred to Committee on Commerce, Science, and Transportation.

S. 2880 (Gillibrand). Rural Broadband Initiative Act of 2009. Establishes an Office of Rural Broadband Initiatives in the Department of Agriculture which would administer the RUS broadband loan and grant programs, and would develop a comprehensive rural broadband strategy. Establishes a National Rural Broadband Innovation Fund, authorized at $20 million for each of fiscal years 2008 through 2012, that would fund experimental and pilot rural broadband projects. Introduced December 14, 2009; referred to Committee on Agriculture, Nutrition, and Forestry.

S. 3110 (Klobuchar). Broadband Service Consumer Protection Act. Seeks to improve consumer protection for purchasers of broadband services by requiring consistent use of broadband service terminology by providers, and requiring clear and conspicuous disclosure to consumers about the actual broadband speed that may reasonably be expected. Introduced March 15, 2010; referred to Committee on Commerce, Science, and Transportation.

S. 3506 (Landrieu). Small Business Broadband and Emerging Information Technology Enhancement Act of 2010. Seeks to improve certain programs of the Small Business Administration to better assist small business customers in accessing broadband technology. Introduced June 17, 2010; referred to Committee on Small Business and Entrepreneurship.

S. 3606 (Kohl). Agriculture, Rural Development, Food and Drug Administration, and Related Agencies Appropriations Act, 2011. For Rural Utilities Service, U.S. Department of Agriculture, provides $22.3 million to support a loan level of $400 million for the broadband loan program, and $17.97 million for broadband community connect grants. Introduced July 15, 2010; referred to Committee on Appropriations. Reported by Committee on Appropriations July 15, 2010 (S.Rept. 111-221) and placed on Senate Legislative Calendar.

S. 3636 (Mikulski). Commerce, Justice, Science, and Related Agencies Appropriations Act, 2011. For FY2011, provides $16 million to NTIA for the administration of BTOP grants and for the development and maintenance of the national broadband map. Introduced July 22, 2010. Reported (S.Rept. 111-229) by Committee on Appropriations July 22, 2010, and placed on Senate Legislative Calendar.

S. 3710 (Murray). Broadband Program Reauthorization Act of 2010. Extends authorization for broadband stimulus programs (BTOP and BIP) at $2 billion each for FY2011 and at such sums as may be necessary for each fiscal year thereafter. Introduced August 5, 2010; referred to Committee on Finance.

S. 3787 (Gillibrand). Upstate Works Act. Provides tax credits to expand broadband service in rural areas. Introduced September 15, 2010; referred to Committee on Finance.

S. 3967 (Landrieu). Small Business Investment and Innovation Act of 2010. Establishes a broadband and emerging information technology coordinator at the Small Business Administration. Introduced November 18, 2010; referred to Committee on Small Business and Entrepreneurship.

S. 3995 (Snowe). Federal Wi-Net Act. Directs the Administrator of the General Services Administration to install Wi-Fi hotspots and wireless neutral host systems in all federal buildings. Introduced December 1, 2010; referred to Committee on Environment and Public Works.

LEGISLATION IN THE 112TH CONGRESS

The 112th Congress is likely to examine the efficacy of federal broadband assistance programs and how they may fit into the context of a national broadband policy. The following is a listing of broadband legislation directly related to the issue of federal assistance for broadband deployment in unserved areas.[53]

*P.L. 112-10 (H.R. 1473)***.** Department of Defense and Full-Year Continuing Appropriations Act, 2011. Rescinds existing unobligated past-year funding for the Rural Broadband Access Loan and Loan Guarantee Program and the Community Connect Grants at the Rural Utilities Service. For FY2011, appropriates $22.3 million to the Rural Broadband Access Loan and Loan Guarantee Program for the cost of broadband loans, and $13.4 million to Community Connect Grants. Signed by President, April 15, 2011.

*H.R. 1083 (Owens)***.** Rural Broadband Initiative Act. Establishes an Office of Rural Broadband Initiatives in the Department of Agriculture which would administer the RUS broadband loan and grant programs, and would develop a comprehensive rural broadband strategy. Introduced March 15, 2011; referred to Committee on Agriculture and in addition to the Committee on Energy and Commerce.

H.R. 1343 (Bass). To return unused or reclaimed funds made available for broadband awards in the American Recovery and Reinvestment Act of 2009 to the Treasury of the United States.

Introduced April 4, 2011; referred to Committee on Energy and Commerce and to Committee on Agriculture. Reported (amended) by the Committee on Energy and Commerce (H.Rept. 112-228) on September 29, 2011. Passed House October 5, 2011. Referred to Senate Committee on Commerce, Science and Transportation October 6, 2011.

H.R. 1695 (Eshoo). Broadband Conduit Deployment Act of 2011. Directs the Secretary of Transportation to require that broadband conduit be installed as part of certain highway construction projects. Introduced May 3, 2011; referred to Committee on Transportation and Infrastructure.

H.R. 2112 (Kingston). Agriculture, Rural Development, Food and Drug Administration, and Related Agencies Appropriations Act, 2012. Provides appropriation for Rural Utilities Service broadband loan program and broadband community connect grants. Introduced June 3, 2011; referred to Committee on Appropriations. Reported by Committee on Appropriations June 3, 2011 (H.Rept. 112-101). Passed House June 16, 2011. Reported by Senate Appropriations Committee September 7, 2011 (S.Rept. 112-73).

*H.R. 2163 (Matsui)***.** Broadband Affordability Act of 2011. Amends the Communications Act of 1934 to establish a Lifeline Assistance Program for universal broadband adoption. Introduced June 14, 2011; referred to Committee on Energy and Commerce.

H.R. 3118 (Farenthold). Lifeline Reform Act. Directs the Federal Communications Commission to revisit the universal service support program under section 254 of the Communications Act of 1934 to reduce waste, fraud, and abuse, and for other purposes.

S. 257 (Landrieu). Small Business Broadband and Emerging Information Technology Enhancement Act of 2011. Seeks to improve certain programs of the Small Business Administration to better assist small business customers in accessing broadband technology. Introduced February 2, 2011; referred to Committee on Small Business and Entrepreneurship.

S. 1659 (Ayotte). To return unused or reclaimed funds made available for broadband awards in the American Recovery and Reinvestment Act of 2009 to the Treasury of the United States. Introduced October 5, 2011; referred to Committee on Commerce, Science and Transportation.

CONCLUSION

To the extent that the 112th Congress may consider various options for encouraging broadband deployment and adoption, a key issue is how to strike a balance between providing federal assistance for unserved and underserved areas where the private sector may not be providing acceptable levels of broadband service, while at the same time minimizing any deleterious effects that government intervention in the marketplace may have on competition and private sector investment.

In addition to loans, loan guarantees, and grants for broadband infrastructure deployment, a wide array of policy instruments are available to policymakers, including universal service reform, tax incentives to encourage private sector deployment, broadband bonds, demand-side incentives (such as assistance to low income families for purchasing computers), regulatory and deregulatory measures, and spectrum policy to spur roll-out of wireless broadband services.

In assessing federal incentives for broadband deployment, the 112th Congress may consider the appropriate mix of broadband deployment incentives to create jobs in the short and long term, the extent to which incentives should target next-generation broadband technologies, the extent to which "underserved" areas with existing broadband providers should receive federal assistance, and whether broadband stimulus projects are being efficiently managed and how they may fit into the context of overall goals for a national broadband policy.

Table 3. Selected Federal Domestic Assistance Programs Related to Broadband and Telecommunications Development

Program	Agency	Description	Funding Amount(FY2009 unless otherwise noted)	Web Links
Broadband Infrastructure Deployment Programs				
Broadband Technology Opportunities Program(BTOP)	National Telecommunications and Information Administration, Dept. of Commerce	Provides competitive grants to public and private sector entities in order to: provide broadband access in unserved and underserved areas; provide broadband support and services to strategic institutions; improve broadband access by public safety agencies; and stimulate broadband demand, economic growth, and job creation.	$4.35 billion (ARRA, P.L. 111-5)	http://www.ntia.doc.gov/ broadbandgrants/
Broadband Initiatives Program (BIP)	Rural Utilities Service, U.S. Dept. of Agriculture	Provides competitive grants, loans, and loan/grant combinations to public and private sector entities in order to provide broadband access in unserved and underserved rural areas.	$2.5 billion for the cost of loans, grants, and loan/grant combinations (ARRA, P.L. 111-5)	http://www.usda.gov/rus/teleco m/arrabroadband. htm
Rural Broadband Access Loan and Loan Guarantee Program	Rural Utilities Service, U.S. Dept. of Agriculture	Provides loan and loan guarantees for facilities and equipment providing broadband service in rural communities	$400 million for cost of money loans	http://www.usda.gov/rus/teleco m/broadband.htm
Community Connect Broadband Grants	Rural Utilities Service, U.S. Dept. of Agriculture	Provides grants to applicants proposing to provide broadband service on a "community-oriented	$13.4 million	http://www.usda.gov/rus/teleco m/index.htm

Program	Agency	Description	Funding Amount(FY2009 unless otherwise noted)	Web Links
		connectivity" basis to rural communities of under 20,000 inhabitants.		
Telecommunications Infrastructure Deployment Programs				
Rural Telephone Loans and Loan Guarantees	Rural Utilities Service, U.S. Dept. of Agriculture	Provides long-term direct and guaranteed loans to qualified organizations for the purpose of financing the improvement, expansion, construction, acquisition, and operation of telephone lines, facilities, or systems to furnish and improve telecommunications service in rural areas	$145 million for hardship loans; $250 million for cost of money loans; and $295 million for FFB Treasury loans	http://www.usda.gov/rus/telecom/index.htm
Distance Learning and Telemedicine Loans and Grants	Rural Utilities Service, U.S. Dept. of Agriculture	Provides seed money to rural community facilities (e.g., schools, libraries, hospitals) for advanced telecommunications systems that can provide health care and educational benefits to rural areas	$34.7 million	http://www.usda.gov/rus/telecom/dlt/dlt.htm
Universal Service High Cost Program	Federal CommunicationsCommission	Provides funding to eligible telecommunications carriers to help pay for telecommunications services in high-cost, rural, and insular areas so that prices charged to customers are reasonably comparable across all regions of the nation.	$4.5 billion (Calendar Year 2008)	http://www.usac.org/hc/

Table 3. (Continued)

Program	Agency	Description	Funding Amount(FY2009 unless otherwise noted)	Web Links
Universal Service Schools and Libraries Program (i.e., E-rate)	Federal Communications Commission	Provides discounts for affordable telecommunications and Internet access services to ensure that schools and libraries have access to affordable telecommunications and information services.	$1.8 billion (Calendar Year 2008)	http://www.universalservice.org/sl/
Universal Service Rural Health Care Pilot Program	Federal CommunicationsCommission	Provides funds to cover 85% of the cost of constructing statewide or regional broadband telehealth networks and of connecting those projects to dedicated nationwide broadband telehealth networks and the public Internet.	$13.05 million committed for funding year 2008 (July 1 to June 30)	http://www.usac.org/rhc-pilot-program/
Appalachian Area Development Program	Appalachian Regional Commission	Project grants to support self-sustaining economic development in the region's most distressed counties and areas. Includes funds for a Telecommunications Initiative involving projects that enable communities to capitalize on broadband access.	$66 million	http://www.arc.gov/index.do?nodeId=21
Delta Area Economic Development	Delta Regional Authority	Grants for self-sustaining economic development projects of eight states in Mississippi Delta region.	$9 million	http://www.dra.gov/state-grant-funding/
Investments for Public Works and	Economic Development	Provides funding for construction of	$129 million	http://www.eda.gov/PDF/FY09

Program	Agency	Description	Funding Amount(FY2009 unless otherwise noted)	Web Links
Economic Development Facilities	Administration, Dept. of Commerce	infrastructure in areas that are not attractive to private investment; most funding is for water and sewer infrastructure but some has been designated for telecommunications and broadband projects.		-EDAPFFO-FINAL.pdf
Library Services and Technology Act Grants to States	Institute of Museum and Library Services, National Foundation on the Arts and the Humanities	Provides funds for a wide range of library services including installation of fiber and wireless networks that provide access to library resources and services.	$171 million	http://www.imls.gov/programs/programs.shtm
Native American and Native Hawaiian Library Services	Institute of Museum and Library Services, National Foundation on the Arts and the Humanities	Grants to support library services including electronically linking libraries to networks.	$3.7 million	http://www.imls.gov/applicants/grants/nativeAmerican.shtm
Programs Related to Applications of Broadband or Telecommunications Technology				
Education Technology State Grants	Office of Elementary and Secondary Education, Dept. of Education	Grants to State Education Agencies for development of information technology to improve teaching and learning in schools.	$269 million	http://www.ed.gov/about/contacts/state/technology.html
Ready to Teach	Office of Assistant Secretary for Educational Research and Improvement, Dept. of Education	Grants for a national telecommunicationbasedprogram to improve the teaching in core curriculum areas.	$10.7 million	http://www.ed.gov/programs/readyteach/index.html

Table 3. (Continued)

Program	Agency	Description	Funding Amount(FY2009 unless otherwise noted)	Web Links
Special Education—Technology and Media Services for Individuals with Disabilities	Office of Special Education and Rehabilitative Services, Dept. of Education	Supports development and application of technology and education media activities for disabled children and adults	$31 million	http://www.ed.gov/about/offices/list/osers/index.html?src=mr/
Telehealth Network Grants	Health Resources and Services Administration, Department of Health and Human Services	Grants to develop sustainable telehealth programs and networks in rural and frontier areas, and in medically unserved areas and populations.	$4 million	http://www.hrsa.gov/telehealth/
Telehealth Resource Center Grant Program	Health Resources and Services Administration, Department of Health and Human Services	Provides grants that support establishment and development of telehealth resource centers to assist health care providers in the development of telehealth services, including decisions regarding the purchase of advanced telecommunications services.	$1.8 million	http://www.hrsa.gov/telehealth/
Licensure Portability Grant Program	Health Resources and Services Administration, Department of Health and Human Services	Provides support for state professional licensing boards to develop and implement state policies that will reduce statutory and regulatory barriers to telemedicine.	$0.7 million plus an estimated $1.5 million under ARRA (P.L. 111-5)	http://www.hrsa.gov/telehealth/
Medical Library Assistance	National Library of Medicine, National Institutes of Health,	Provides funds to train professional personnel; strengthen library and	$66 million	http://www.nlm.nih.gov/ep/extramural.html

Program	Agency	Description	Funding Amount(FY2009 unless otherwise noted)	Web Links
	Department of Health and Human Services	information services; facilitate access to and delivery of health science information; plan and develop advanced information networks; support certain kinds of biomedical publications; and conduct research in medical informatics and related sciences.		
Smart Grid Investment Grant Program and Smart Grid Demonstration Program	Department of Energy	Provides support for modernizing the electric grid, which likely includes some broadband for Smart Grid.	$3.9 billion (ARRA, P.L. 111-5)	http://www.energy.gov/news2009/7503.htm
National Environmental Information Exchange Network Grant Program	Environmental Protection Agency	Provides funding to states, territories, and federally recognized Indian Tribes to support the development of an Environmental Information Exchange Network, including broadband infrastructure.	$12 million	http://epa.gov/exchangenetwork/grants/

Source: Compiled by CRS based on GAO Report, Broadband Deployment Plan Should Include Performance Goals and Measures To Guide Federal Investment, May 2009, Tables 2 and 3; FCC Report, Bringing Broadband to Rural America, May 22, 2009, Appendix B; and the Catalog of Federal Domestic Assistance.

End Notes

[1] The term "digital divide" can also refer to international disparities in access to information technology. This report focuses on domestic issues only.

[2] FCC, *Internet Access Services: Status as of December 30, 2010*, released October 2011, p. 16. Available at http://transition.fcc.gov/Daily_Releases/Daily_Business/2011/db1011/DOC-310261A1.pdf.

[3] Ibid.

[4] Initially, and for many years following, the FCC defined broadband (or more specifically "high-speed lines") as over 200 kilobits per second (kbps) in at least one direction, which was roughly four times the speed of conventional dialup Internet access. In recent years, the 200 kbps threshold was considered too low, and on March 19, 2008, the FCC adopted a report and order (FCC 08-89) establishing new categories of broadband speed tiers for data collection purposes. Specifically, 200 kbps to 768 kbps will be considered "first generation," 768 kbps to 1.5 Mbps as "basic broadband tier 1," and increasingly higher speed tiers as broadband tiers 2 through 7 (tier seven is greater than or equal to 100 Mbps in any one direction). Tiers can change as technology advances.

[5] U.S. Department of Commerce, National Telecommunications and Information Administration, *Digital Nation: Expanding Internet Usage*, February 2011, p. 2, available at http://www.ntia.doc.gov/reports/2011/NTIA_Internet_Use_Report_February_2011.pdf.

[6] Federal Communications Commission, *Connecting America: The National Broadband Plan, March 17, 2010*, p. 20.

[7] Data as of June 30, 2010. Available at http://www.broadbandmap.gov/summarize/nationwide.

[8] National Telecommunications and Information Administration, Press Release, "Commerce's NTIA Unveils National Broadband Map and New Broadband Survey Adoption Results," February 17, 2011, available at http://www.ntia.doc.gov/press/2011/National Broadband Map_02172011.html.

[9] Horrigan, John, Federal Communications Commission, *Broadband Adoption and Use in America*, OBI Working Paper Series No. 1, February 2010, p. 5, available at http://hraunfoss.fcc.gov/edocs_public/attachmatch/DOC-296442A1.pdf.

[10] *Digital Nation*, p. 20.

[11] Smith, Aaron, Pew Internet & American Life Project, *Home Broadband 2010*, August 11, 2010, available at http://www.pewinternet.org/~/media//Files/Reports/2010/Home% 20 broadband%202010.pdf.

[12] Federal Communications Commission, *Seventh Broadband Deployment Report*, FCC 11-78, released May 20, 2011, p. 15, available at http://transition.fcc.gov/Daily_Releases/Daily_Business/2011/db0520/FCC-11-78A1.pdf.

[13] For more information on rural broadband and broadband programs at the Rural Utilities Service, see CRS Report RL33816, *Broadband Loan and Grant Programs in the USDA's Rural Utilities Service*, by Lennard G. Kruger.

[14] See for example: National Exchange Carrier Association (NECA), Trends 2006: Making Progress With Broadband, 2006, 26 p. Available at http://www.neca.org/media/trends_brochure_website.pdf.

[15] *Digital Nation*, p. 16.

[16] *Home Broadband 2010*, p. 8.

[17] NTIA, National Broadband Map, *Broadband Statistics Report: Broadband Availability in Urban vs. Rural Areas*, p. 7, available at http://www.broadbandmap.gov/download/reports/national-broadband-map-broadband-availability-in-rural-vs-urban-areas.pdf.

[18] *Seventh Broadband Deployment Report*, p. 27.

[19] The terrain of rural areas can also be a hindrance to broadband deployment because it is more expensive to deploy broadband technologies in a mountainous or heavily forested area. An additional added cost factor for remote areas can be the expense of "backhaul" (e.g., the "middle mile") which refers to the installation of a dedicated line which transmits a signal to and from an Internet backbone which is typically located in or near an urban area.

[20] Gillett, Sharon E., Massachusetts Institute of Technology, *Measuring Broadband's Economic Impact*, report prepared for the Economic Development Administration, U.S. Department of Commerce, February 28, 2006 p. 4.

[21] Crandall, Robert, William Lehr, and Robert Litan, *The Effects of Broadband Deployment on Output and Employment: A Cross-sectional Analysis of U.S. Data*, June 2007, 20 pp. Available at http://www3.brookings.edu/views/papers/crandall/200706litan.pdf.

[22] Mark Dutz, Jonathan Orszag, and Robert Willig, *The Substantial Consumer Benefits of Broadband Connectivity for U.S. Households*, Internet Innovation Alliance, July 2009, p. 4, http://internetinnovation.org/files/special-reports/CONSUMER_BENEFITS_OF_ BROADBAND.pdf.

[23] OECD, *OECD Broadband Portal*. Available at http://www.oecd.org/sti/ict/broadband.

[24] OECD, Directorate for Science, Technology and Industry, *The Development of Broadband Access in OECD Countries*, October 29, 2001, 63 pp. For a comparison of government broadband policies, also see OECD, Directorate for Science, Technology and Industry, *Broadband Infrastructure Deployment: The Role of Government Assistance*, May 22, 2002, 42 pp.

[25] See Turner, Derek S., Free Press, *Broadband Reality Check II: The Truth Behind America's Digital Divide*, August 2006, pp 8-11. Available at http://www.freepress.net/files/bbrc2-final.pdf; and Turner, Derek S., Free Press, *'Shooting the Messenger' Myth vs. Reality: U.S. Broadband Policy and International Broadband Rankings*, July 2007, 25 pp., available at http://www.freepress.net/files/shooting_the_messenger.pdf.

[26] National Telecommunications and Information Administration, *Fact Sheet: United States Maintains Information and Communication Technology (ICT) Leadership and Economic Strength*, at http://www.ntia.doc.gov/ntiahome/press/2007/ICTleader_042407.html.

[27] See Wallsten, Scott, Progress and Freedom Foundation, *Towards Effective U.S. Broadband Policies*, May 2007, 19 pp. Available at http://www.pff.org/issues-pubs/pops/ pop14.7 usbroadbandpolicy.pdf. Also see Ford, George, Phoenix Center, *The Broadband Performance Index: What Really Drives Broadband Adoption Across the OECD?*, Phoenix Center Policy Paper Number 33, May 2008, 27 pp; available at http://www.phoenix-center.org/pcpp/PCPP33Final.pdf.

[28] See price and services and speed data on OECD Broadband Portal, available at http://www.oecd.org/sti/ict/broadband; Turner, Derek S., Free Press, *Broadband Reality Check II: The Truth Behind America's Digital Divide*, August 2006, pp 5-9; Kende, Michael, Analysis Consulting Limited, *Survey of International Broadband Offerings*, October 4, 2006, 12 p, available at http://www.analysys.com/pdfs/Broadband Performance Survey.pdf; and Atkinson, Robert D., The International Technology and Innovation Foundation, *Explaining International Broadband Leadership*, May 2008, 108 p, available at http://www.itif.org/files/ExplainingBBLeadership.pdf.

[29] Federal Communications Commission, *Notice of Inquiry*, "Concerning the Deployment of Advanced Telecommunications Capability to All Americans in a Reasonable and Timely Fashion, and possible Steps to Accelerate Such Deployment Pursuant to Section 706 of the Telecommunications Act of 1996," FCC 04-55, March 17, 2004, p. 6.

[30] FCC News Release, *FCC Improves Data Collection to Monitor Nationwide Broadband Rollout*, November 9, 2004. Available at http://hraunfoss.fcc.gov/edocs_public/attachmatch/DOC-254115A1.pdf.

[31] Federal Communications Commission, *Notice Proposed Rulemaking*, "Development of Nationwide Broadband Data to Evaluate Reasonable and Timely Deployment of Advanced Services to All Americans, Improvement of Wireless Broadband Subscribership Data, and Development of Data on Interconnected Voice Over Internet Protocol (VoIP) Subscribership," WC Docket No. 07-38, FCC 07-17, released April 16, 2007, 56 pp. 32 FCC, News Release, "FCC Expands, Improves Broadband Data Collection," March 19, 2008. Available at http://hraunfoss.fcc.gov/edocs_public/attachmatch/DOC-280909A1.pdf.

[33] For more information on the national broadband mapping program and the State Broadband Data and Development Program, see http://www.ntia.doc.gov/broadbandgrants/broadband mapping.html.

[34] *Seventh Broadband Deployment Report*, p. 2.

[35] Ibid., p. 94.

[36] Ibid., p. 7-10.

[37] Available at http://www.broadband.gov/plan/. For more information on the National Broadband Plan, see CRS Report R41324, *The National Broadband Plan*, by Lennard G. Kruger et al.

[38] Federal Communications Commission, Connecting America: *The National Broadband Plan*, March 17, 2010, p. 9.

[39] Ibid., p. 11.

[40] Ibid.

[41] Ibid.

[42] The White House, Office of the Press Secretary, "Statement from the President on the National Broadband Plan," March 16, 2010, available at http://www.whitehouse.gov/the-press-office/statement-president-national-broadband-plan.

[43] See CRS Report R40436, *Broadband Infrastructure Programs in the American Recovery and Reinvestment Act*, by Lennard G. Kruger.

[44] The section on universal service was prepared by Angele Gilroy, Specialist in Telecommunications, Resources, Science and Industry Division. For more information on universal service, see CRS Report RL33979, *Universal Service Fund: Background and Options for Reform*, by Angele A. Gilroy.

[45] Communications Act of 1934, As Amended, Title I §1 [47 USC. 151].

[46] Many states participate in or have programs that mirror FCC universal service mechanisms to help promote universal service goals within their states.

[47] Additional FCC policies such as rate averaging and pooling have also been implemented to assist high cost carriers.

[48] For additional information on this program, including funding commitments, see the E-rate website: http://www.universalservice.org/sl/.

[49] For additional information on this program, including funding commitments, see the RHCD website: http://www.universalservice.org/rhc/.

[50] The Joint Board recommended that the definition of those services that qualify for universal service support be expanded and that the nation's communications goals include the universal availability of: mobility services (i.e., wireless voice); broadband Internet services; and voice services at affordable and comparable rates for all rural and non-rural areas. For a copy of this recommendation see http://hraunfoss.fcc. gov/edocs_ public/attachmatch/FCC-07J-4A1.pdf.

[51] For a further discussion of the role of the Universal Service Fund in the national broadband plan and the FCC's actions to redesign the USF to become a vehicle for broadband deployment see CRS Report RL33979, *UniversalService Fund: Background and Options for Reform*, by Angele A. Gilroy.

[52] For information on existing broadband programs at RUS, see CRS Report RL33816, *Broadband Loan and Grant Programs in the USDA's Rural Utilities Service*, by Lennard G. Kruger.

[53] For information on public safety wireless broadband legislation, see CRS Report R41842, *Funding Emergency Communications: Technology and Policy Considerations*, by Linda K. Moore.

INDEX

A

B

C

D

E

F

G

H

I

J

L

M

N

O

P

V

W